Algebra 1

LARSON
BOSWELL
KANOLD
STIFF

Applications • Equations • Graphs

Chapter 9
Resource Book

The Resource Book contains the wide variety of blackline masters available for Chapter 9. The blacklines are organized by lesson. Included are support materials for the teacher as well as practice, activities, applications, and assessment resources.

McDougal Littell
A HOUGHTON MIFFLIN COMPANY
Evanston, Illinois • Boston • Dallas

Contributing Authors

The authors wish to thank the following individuals for their contributions to the Chapter 9 Resource Book.

Rita Browning
Linda E. Byrom
José Castro
Christine A. Hoover
Carolyn Huzinec
Karen Ostaffe
Jessica Pflueger
Barbara L. Power
Joanne Ricci
James G. Rutkowski
Michelle Strager

Pages 20, 65, 107, 123: Excerpted and adapted from The World Book Encyclopedia. Copyright © 2000 World Book, Inc. By Permission of the publisher. www.worldbook.com

ISBN: 0-618-02047-0

15 14 13 12 11 10 - CKI - 06 05 04

Contents

9   *Quadratic Equations and Functions*

Contents

Contents

Descriptions of Resources

This Chapter Resource Book is organized by lessons within the chapter in order to make your planning easier. The following materials are provided:

Tips for New Teachers These teaching notes provide both new and experienced teachers with useful teaching tips for each lesson, including tips about common errors and inclusion.

Parent Guide for Student Success This guide helps parents contribute to student success by providing an overview of the chapter along with questions and activities for parents and students to work on together.

Prerequisite Skills Review Worked-out examples are provided to review the prerequisite skills highlighted on the Study Guide page at the beginning of the chapter. Additional practice is included with each worked-out example.

Strategies for Reading Mathematics The first page teaches reading strategies to be applied to the current chapter and to later chapters. The second page is a visual glossary of key vocabulary.

Lesson Plans and Lesson Plans for Block Scheduling This planning template helps teachers select the materials they will use to teach each lesson from among the variety of materials available for the lesson. The block-scheduling version provides additional information about pacing.

Warm-Up Exercises and Daily Homework Quiz The warm-ups cover prerequisite skills that help prepare students for a given lesson. The quiz assesses students on the content of the previous lesson. (Transparencies also available)

Activity Support Masters These blackline masters make it easier for students to record their work on selected activities in the Student Edition.

Alternative Lesson Openers An engaging alternative for starting each lesson is provided from among these four types: *Application, Activity, Graphing Calculator,* or *Visual Approach.* (Color transparencies also available)

Graphing Calculator Activities with Keystrokes Keystrokes for four models of calculators are provided for each Technology Activity in the Student Edition, along with alternative Graphing Calculator Activities to begin selected lessons.

Practice A, B, and C These exercises offer additional practice for the material in each lesson, including application problems. There are three levels of practice for each lesson: A (basic), B (average), and C (advanced).

Contents

Reteaching with Practice These two pages provide additional instruction, worked-out examples, and practice exercises covering the key concepts and vocabulary in each lesson.

Quick Catch-Up for Absent Students This handy form makes it easy for teachers to let students who have been absent know what to do for homework and which activities or examples were covered in class.

Cooperative Learning Activities These enrichment activities apply the math taught in the lesson in an interesting way that lends itself to group work.

Interdisciplinary Applications/Real-Life Applications Students apply the mathematics covered in each lesson to solve an interesting interdisciplinary or real-life problem.

Math and History Applications This worksheet expands upon the Math and History feature in the Student Edition.

Challenge: Skills and Applications Teachers can use these exercises to enrich or extend each lesson.

Quizzes The quizzes can be used to assess student progress on two or three lessons.

Chapter Review Games and Activities This worksheet offers fun practice at the end of the chapter and provides an alternative way to review the chapter content in preparation for the Chapter Test.

Chapter Tests A, B, and C These are tests that cover the most important skills taught in the chapter. There are three levels of test: A (basic), B (average), and C (advanced).

SAT/ACT Chapter Test This test also covers the most important skills taught in the chapter, but questions are in multiple-choice and quantitative-comparison format. (See *Alternative Assessment* for multi-step problems.)

Alternative Assessment with Rubrics and Math Journal A journal exercise has students write about the mathematics in the chapter. A multi-step problem has students apply a variety of skills from the chapter and explain their reasoning. Solutions and a 4-point rubric are included.

Project with Rubric The project allows students to delve more deeply into a problem that applies the mathematics of the chapter. Teacher's notes and a 4-point rubric are included.

Cumulative Review These practice pages help students maintain skills from the current chapter and preceding chapters.

Tips for New Teachers

For use with Chapter 9

LESSON 9.1

COMMON ERROR Some students might think that they can find the square root of a negative number, by doing something like $\sqrt{-4} = -2$. Show students that this is not true by using the definition of the square root of a number; in other words, by showing them that $(-2)^2 \neq -4$. You might have to review the different sets of numbers students already know to explain to them what not having *real* square roots means.

TEACHING TIP Ask your students to create a list of the first fifteen perfect squares that are also whole numbers. Ideally they should memorize this list, but they can also refer to it as needed. Knowing these values will help students estimate square roots *without* a calculator. For instance, $\sqrt{150}$ must be between 12 and 13, because $12^2 = 144$ and $13^2 = 169$. Since 150 is much closer to 144 than to 169, an initial estimate could be $\sqrt{150} \approx 12.2$, closer to 12 than to 13.

COMMON ERROR When solving quadratic equations in which $b = 0$, some students only take the positive square root. Their answer will check in the original equation, but these students completely forget about a second answer: the negative square root. You can help them to remember to take both square roots by using the $\pm$ symbol in front of the radical whenever you take the square roots of each side of this type of equation. In addition, when solving a word problem such as Example 7 on page 506, always write both mathematical answers to the problem ($t \approx \pm 1.4$ for this example). Then ask students to write a sentence explaining whether these answers make sense for the specific problem they are working on (you can even link this topic to the ideas of domain and range).

LESSON 9.2

COMMON ERROR Students might make errors such as

$$\frac{\sqrt{15}}{3} = \sqrt{5}.$$

Ask these students to evaluate both sides of their equation with a calculator, so that they can see their mistake. Then remind students that they can reduce fractions *inside* the radical or *outside* the radical, but they cannot mix these two.

COMMON ERROR Students might write $\sqrt{50} = 25\sqrt{2}$ or $\sqrt{50} = 2\sqrt{5}$. Typically, these mistakes are made by students who try to solve these problems without showing any steps. Emphasize the need to write down another step where the original radical is shown as the product of a perfect square and another number ($\sqrt{50} = \sqrt{25 \cdot 2}$). Always write the perfect square first, to help students see what remains inside the radical after they take the square root of the perfect square.

LESSON 9.3

TEACHING TIP Show your students that c is always the value of the y-intercept for a quadratic function of the form $y = ax^2 + bx + c$, because when $x = 0$, $y = c$. Students can quickly sketch graphs for quadratic equations if they remember this fact and know whether the parabola opens up or down and how to find its vertex. This can be a valuable skill for multiple choice standardized tests, where students typically need to match an equation and its graph.

LESSON 9.4

TEACHING TIP The connection between the solutions of a quadratic equation and the x-intercepts of its related function must be made clear to the students. Point out that if we make $y = 0$ in a quadratic function, we get the related quadratic equation. Finding the solutions to that equation is the same as finding the values of the function where the y-coordinate is 0. Since we know that such points are always on the x-axis, what we are looking for are the x-intercepts.

TEACHING TIP Ask students to solve a quadratic equation where the solutions are not integers, such as $6x^2 + x - 2 = 0$ (solutions are $\frac{1}{2}$, $-\frac{8}{3}$). They will not be able to find the exact points where the graph crosses the x-axis. Instead, they will have to estimate the answers (they should be able to determine between which two integers the solution lies).

Tips for New Teachers

For use with Chapter 9

LESSON 9.4 (CONT.)

TEACHING TIP Ask your students to solve some quadratic equations where the related function crosses the *x*-axis in just one point or does not cross it at all. Discuss how many solutions these equations have and how we can tell from the graph of the related function how many solutions the equation has. You can continue this topic when you cover the discriminant in Lesson 9.6.

LESSON 9.5

TEACHING TIP Start the class by either showing or reviewing a quadratic equation where the solutions are not integers. Since graphing the related function will not give us the exact answers for this equation, we need an alternate method to find the solutions. This is when the quadratic formula is useful.

COMMON ERROR Students can make all kinds of mistakes using the quadratic formula, so you might want to spend some time making sure that they are able to get the correct answers. First, remind students that the equation must be in standard form so that they can identify the values of *a*, *b*, and *c*. Second, review the meaning of "$-b$" as the opposite of *b*, so that students do not think of that term as always negative. Finally, beware of those students who incorrectly use their calculator to evaluate the last step of the formula. They might be calculating

$$-9 \pm \frac{5}{2} \text{ instead of } \frac{-9 \pm 5}{2}.$$

Review order of operations and remind these students that they need to put parentheses around the numerator if they want to find the answer with their calculator.

LESSON 9.6

INCLUSION Many students have never heard the word *discriminant*, but most of them know what *discriminate* means. Show the relationship between these two words as a means to explain the role of the discriminant in the quadratic formula. You can tell your students that the quadratic formula "*discriminates* against negative *discriminants*."

LESSON 9.7

TEACHING TIP The process followed to graph quadratic inequalities is basically the same as the one used to graph linear inequalities. You might want to start the class by reviewing how to graph linear inequalities and then ask students what the process would be for quadratic inequalities.

LESSON 9.8

TEACHING TIP You might want to include in your lesson an example that can be modeled by using a linear equation to review how to graph and find the equation of the best-fitting line.

TEACHING TIP Remind students that the objective of a model is to re-create the data as closely as possible, but there can be some discrepancies between the original set of data and the one generated by the model. These discrepancies might not even be due to the model, but rather to the method used to collect data or the nature of the data itself.

Outside Resources

BOOKS/PERIODICALS
Mercer, Joseph. "Teaching Graphing Concepts with Graphics Calculators." *Mathematics Teacher* (April 1995); pp. 268–273.

ACTIVITIES/MANIPULATIVES
Algeblocks. Three dimensional blocks representing variables and units. Vernon Hills, IL; ETA.

SOFTWARE
Harvey, Wayne and Judah L. Schwartz. *Visualizing Algebra : The Function Analyzer*. Pleasantville, NY; Sunburst Communications.

VIDEOS
Apostel, Tom. *Polynomials*. Reston, VA; NCTM.

NAME _____ DATE _____

Parent Guide for Student Success

For use with Chapter 9

Chapter Overview One way that you can help your student succeed in Chapter 9 is by discussing the lesson goals in the chart below. When a lesson is completed, ask your student to interpret the lesson goals for you and to explain how the mathematics of the lesson relates to one of the key applications listed in the chart.

Lesson Title	Lesson Goals	Key Applications
9.1: Solving Quadratic Equations by Finding Square Roots	Evaluate and approximate square roots. Solve a quadratic equation by finding square roots.	• Engineering • Computer Sales • Minerals
9.2: Simplifying Radicals	Use properties of radicals to simplify radicals. Use quadratic equations to model real-life problems.	• Boat Racing • Body Surface Area • Tsunami
9.3: Graphing Quadratic Functions	Sketch the graph of a quadratic function. Use quadratic models in real-life settings.	• Track and Field • Water Arc • Gold Production
9.4: Solving Quadratic Equations by Graphing	Solve a quadratic equation graphically. Use quadratic models in real-life settings.	• Shot Put • Swiss Cheese • RV Sales
9.5: Solving Quadratic Equations by the Quadratic Formula	Use the quadratic formula to solve a quadratic equation. Use quadratic models in real-life settings.	• Balloon Competition • Peregrine Falcon • Baseball
9.6: Applications of the Discriminant	Use the discriminant to find the number of solutions of a quadratic equation. Apply the discriminant to solve real-life problems.	• Camping • Government Payroll • Financial Analysis
9.7: Graphing Quadratic Inequalities	Sketch the graph of a quadratic inequality. Use quadratic inequalities as real-life models.	• Bridge Building • Diamonds • Milk Consumption
9.8: Comparing Linear, Exponential, and Quadratic Models	Choose a model that best fits a collection of data and use models in real-life settings.	• Chambered Nautilus • Pendulums • NHL Attendance

Study Strategy

Explaining Ideas is the study strategy featured in Chapter 9 (see page 502). Having your student try to explain material to you can provide an opportunity for him or her to pull ideas together, to identify and overcome misunderstandings, and to review and prepare for tests.

NAME _____ DATE _____

Parent Guide for Student Success

For use with Chapter 9

Key Ideas Your student can demonstrate understanding of key concepts by working through the following exercises with you.

Lesson	Exercise
9.1	Solve the equation. Write the solutions as integers or as radical expressions. $11x^2 - 14 = 30$
9.2	Simplify the expression. $\left(\sqrt{12} \cdot \sqrt{25}\right)\sqrt{3}$
9.3	You throw a baseball whose path can be modeled by $h = -16t^2 + 8t + 4$, where h (in feet) is the height of the baseball t seconds after it is released. How long does it take the baseball to reach its highest point? What is the maximum height of the baseball?
9.4	Use a graphing calculator to estimate how many seconds it will take the baseball in Exercise 9.3 to hit the ground.
9.5	Use the quadratic formula to solve the equation. $2x^2 - 3x - 5 = 0$
9.6	Evaluate the discriminant of $3x^2 - 5x + 4 = 0$. How many solutions does the equation have? Does the graph of $y = 3x^2 - 5x + 4$ cross the x-axis?
9.7	Use the graph of the inequality $y \le -x^2 + 5$ to determine whether $(-2, 2)$ is a solution of the inequality.
9.8	A quilting club worked on a quilt for 4 days in a row. The size of the quilt after Day 1 was 63 in.^{2} After Day 2, it was 112 in.^{2}, and after Day 3, it was 175 in.^{2} If the club continued in the same pattern, how big was the quilt after Day 4?

Home Involvement Activity

Directions: Design a rectangular garden with a sidewalk border all around the outer edge. Decide the overall dimensions of your garden with the sidewalk and then find the area. Take 25% of the area and let this be the area of the sidewalk. Find the remaining garden area. Let x be the width of the sidewalk. Write and solve a quadratic equation to find what width you should make the sidewalk border.

Answers
9.1: 2, −2 9.2: 30 9.3: 0.25 sec; 5 ft 9.4: about 0.81 sec 9.5: $\frac{5}{2}$, − 1 9.6: −23; none; no 9.7: not a solution 9.8 252 in.^{2}

NAME _____ DATE _____

Prerequisite Skills Review

For use before Chapter 9

EXAMPLE 1 *Evaluating a Variable Expression*

Evaluate the expression.

$(-8.5 \cdot x)(-3)$ when $x = -2$

SOLUTION

$(-8.5 \cdot x)(-3) = [-8.5 \cdot (-2)] \cdot (-3)$	Substitute -2 for x first.
$= -8.5 \cdot [(-2) \cdot (-3)]$	Use associative property.
$= -8.5 \cdot 6$	$-2 \cdot (-3) = 6$
$= -51$	Simplify.

Exercises for Example 1

Evaluate the expression.

1. $200 - 5y^3$ when $y = -3$ **2.** $6x^2 \div \frac{5}{9}$ when $x = -5$

3. $-\dfrac{4a}{a + b}$ when $a = 10$ and $b = -4$ **4.** $\dfrac{s^2 + 11}{9 + t}$ when $s = 7$ and $t = 21$

EXAMPLE 2 *Graphing an Equation*

Use a table of values to graph the equation.

$3y + 6 = 4x$

SOLUTION

Rewrite the equation in function form by solving for y.

$3y + 6 = 4x$	Write original equation.
$3y = 4x - 6$	Subtract 6 from each side.
$y = \frac{4}{3}x - 2$	Divide each side by 3.

Choose a few values of x and made a table of values.

Choose x	Substitute to find the corresponding y-values
-2	$y = \frac{4}{3}(-2) - 2 = -\frac{14}{3} = -4\frac{2}{3}$
-1	$y = \frac{4}{3}(-1) - 2 = -\frac{10}{3} = -3\frac{1}{3}$
0	$y = \frac{4}{3}(0) - 2 = -2$
1	$y = \frac{4}{3}(1) - 2 = -\frac{2}{3}$
2	$y = \frac{4}{3}(2) - 2 = \frac{2}{3}$

NAME _____ DATE _____

Prerequisite Skills Review

For use before Chapter 9

With this table of values you have found five
solutions

$$\left(-2, -4\tfrac{2}{3}\right), \left(-1, -3\tfrac{1}{3}\right), (0, -2), \left(1, -\tfrac{2}{3}\right), \left(2, \tfrac{2}{3}\right)$$

Plot the points. You can see that they all lie on a line. The
graph through the points is the graph of the equation.

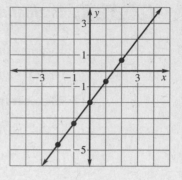

Exercises for Example 2

Use a table of values to graph the equation.

5. $y = \tfrac{4}{5}x + 15$

6. $y - 5 = -8x + 9$

7. $x + 7y = 28$

8. $6x + 3y = 14$

EXAMPLE 3 *Checking solutions of a Linear Inequality*

Check whether the ordered pair is a solution.

$\tfrac{5}{2}x - \tfrac{9}{2}y \geq 10, (-2, 4)$

SOLUTION

$\tfrac{5}{2}x - \tfrac{9}{2}y \geq 10, (-2, 4)$	Write original equation.
$\tfrac{5}{2}(-2) - \tfrac{9}{2}(4) \geq 10$	Substitute -2 in for x and 4 in for y.
$-\tfrac{10}{2} - \tfrac{36}{2} \geq 10$	Simplify.
$-5 - 18 \geq 10$	
$-23 \geq 10$	Not true.

$(-2, 4)$ is not a solution of the linear inequality.

Exercises for Example 3

Check whether the ordered pair is a solution.

9. $9y - 2x > -5, (-3, 9)$

10. $3.5x \geq 13 - 0.12y, (0.23, 9)$

11. $x - \tfrac{5}{8} \leq \tfrac{2}{3}(y - 1), \left(\tfrac{9}{8}, \tfrac{5}{3}\right)$

12. $4x - 12 < \tfrac{1}{3}(1 - 16x), (-9, 7)$

NAME _____ DATE _____

Strategies for Reading Mathematics

For use with Chapter 9

Strategy: Reading Tables

In some cases, you may prefer to find squares and square roots using the Table of Squares and Square Roots on page 811 of your text. (Note that the table gives square roots rounded to three decimal places.) Many tables, including this one, have several columns of data for each number. It is important that you always check the column heads as you read numbers in a table to make sure you have found the data you need.

Column heads list the content of each column.

Read across each row. →
Then read the column head to check your place.

No.	Square	Sq. Root	No.	Square	Sq. Root
1	1	1.000	51	2601	7.141
2	4	1.414	52	2704	7.211
3	9	1.732	53	2809	7.280
4	16	2.000	54	2916	7.348
5	25	2.236	55	3025	7.416
6	36	2.449	56	3136	7.483
7	49	2.646	57	3249	7.550
8	64	2.828	58	3364	7.616
9	81	3.000	59	3481	7.681
10	100	3.162	60	3600	7.746

Think about the meaning of each number.

$$100 = 10^2 \qquad 3.162 = \sqrt{10}$$

STUDY TIP
Use a Ruler

A straight edge, such as a ruler or the edge of a piece of paper, can help you keep your place as you read a table. Placing the ruler below the row you are reading helps focus your eyes on the entry you need to read.

STUDY TIP
Estimating from the Table

The Table of Squares and Square Roots may be used to estimate certain squares and square roots. For example, you can determine that $7.6^2 \approx 58$ and $\sqrt{3030} \approx 55$. Unless a rough estimate is sufficient, it is generally better to use a calculator for such numbers.

Questions

Use the Table of Squares and Square Roots above for Questions 1–3.

1. Find the square of each integer.

   a. 52   **b.** 60   **c.** 53   **d.** 57

2. Find the square root of each integer.

   a. 3   **b.** 6   **c.** 51   **d.** 58

3. Estimate the value of each expression to the nearest whole number.

   a. 3.2^2 **b.** 1.8^2 **c.** $\sqrt{65}$ **d.** $\sqrt{3475}$

Chapter Support

Strategies for Reading Mathematics

For use with Chapter 9

Visual Glossary

The Study Guide on page 502 lists the key vocabulary for Chapter 9 as well as review vocabulary from previous chapters. Use the page references on page 502 or the Glossary in the textbook to review key terms from prior chapters. Use the visual glossary below to help you understand some of the key vocabulary in Chapter 9. You may want to copy these diagrams into your notebook and refer to them as you complete the chapter.

GLOSSARY

quadratic equation (p. 505) An equation that can be written in the form $ax^2 + bx + c = 0$, where $a \neq 0$.

discriminant (p. 541) The expression $b^2 - 4ac$ where a, b, and c are coefficients of the quadratic equation $ax^2 + bx + c = 0$.

radical expression (p. 504) An expression that involves square roots.

roots of a quadratic equation (p. 526) The solutions of $ax^2 + bx + c = 0$.

parabola (p. 518) The U-shaped graph of a quadratic function.

quadratic inequality (p. 548) An inequality that can be written as follows:

$y < ax^2 + bx + c$

$y \leq ax^2 + bx + c$

$y > ax^2 + bx + c$

$y \geq ax^2 + bx + c$

graph of a quadratic inequality (p. 548) The graph of all ordered pairs (x, y) that are solutions of the inequality.

Solving Quadratic Equations

As you solve quadratic equations, you will apply new vocabulary in many steps in the process.

$0 = x^2 + 2x - 3$ ◄———— quadratic equation

$x = \dfrac{-b \pm \sqrt{b^2 - 4ac}}{2a}$ ◄———— discriminant

$= \dfrac{-2 \pm \boxed{\sqrt{4 - 4(1)(-3)}}}{2(1)}$ ◄———— radical expression

$= \dfrac{-2 \pm \sqrt{16}}{2} = \dfrac{-2 \pm 4}{2}$

$x = -1$ or $x = -3$ ◄———— roots

Graphing Quadratic Inequalities

To sketch the graph of a quadratic inequality, first graph the related quadratic equation. The graph of the equation is a parabola. Test a point not on the parabola to determine which region to shade.

To graph $y < -x^2 + 2x + 2$, graph $y = -x^2 + 2x + 2$. First, find the vertex.

$x = \dfrac{-b}{2a} = \dfrac{-2}{2(-1)} = 1 \quad y = -(1)^2 + 2(1) + 2 = 3 \quad \text{vertex} = (1, 3)$

Make a table of values:

x	-2	-1	0	1	2	3	4
y	-6	-1	2	3	2	-1	-6

Test a point:

$0 < -0^2 + 2(0) + 2$,

so $(0, 0)$ is a solution

parabola

graph of the quadratic inequality

Algebra 1
Chapter 9 Resource Book

Lesson Plan

1-day lesson (See *Pacing the Chapter,* TE pages 500C–500D)   **For use with pages 503–510**

GOALS
1. **Evaluate and approximate square roots.**
2. **Solve a quadratic equation by finding square roots.**

State/Local Objectives _____

_____

✓ **Check the items you wish to use for this lesson.**

STARTING OPTIONS
____ Prerequisite Skills Review: CRB pages 5–6
____ Strategies for Reading Mathematics: CRB pages 7–8
____ Warm-Up or Daily Homework Quiz: TE pages 503 and 490, CRB page 11, or Transparencies

TEACHING OPTIONS
____ Motivating the Lesson: TE page 504
____ Lesson Opener (Application): CRB page 12 or Transparencies
____ Graphing Calculator Activity with Keystrokes: CRB page 13
____ Examples 1–7: SE pages 503–506
____ Extra Examples: TE pages 504–506 or Transparencies
____ Closure Question: TE page 506
____ Guided Practice Exercises: SE page 507

APPLY/HOMEWORK
Homework Assignment
____ Basic 23, 26, 29, and every 3rd Ex. through 77, 79–81, 95, 96, 104, 107, 110, 111
____ Average 23, 26, 29, and every 3rd Ex. through 77, 79–81, 87, 95, 96, 104, 107, 110, 111
____ Advanced 23, 26, 29, and every 3rd Ex. through 77, 78–81, 87, 95–100, 104, 107, 110, 111

Reteaching the Lesson
____ Practice Masters: CRB pages 14–16 (Level A, Level B, Level C)
____ Reteaching with Practice: CRB pages 17–18 or Practice Workbook with Examples
____ Personal Student Tutor

Extending the Lesson
____ Applications (Interdisciplinary): CRB page 20
____ Challenge: SE page 510; CRB page 21 or Internet

ASSESSMENT OPTIONS
____ Checkpoint Exercises: TE pages 504–506 or Transparencies
____ Daily Homework Quiz (9.1): TE page 510, CRB page 24, or Transparencies
____ Standardized Test Practice: SE page 509; TE page 510; STP Workbook; Transparencies

Notes _____

_____

_____

TEACHER'S NAME _____ CLASS _____ ROOM _____ DATE _____

Lesson Plan for Block Scheduling

Half-day lesson (See *Pacing the Chapter,* TE pages 500C–500D)   For use with pages 503–510

GOALS 1. **Evaluate and approximate square roots.**
2. **Solve a quadratic equation by finding square roots.**

State/Local Objectives _____

_____

_____ .

✓ **Check the items you wish to use for this lesson.**

STARTING OPTIONS

____ Prerequisite Skills Review: CRB pages 5–6
____ Strategies for Reading Mathematics: CRB pages 7–8
____ Warm-Up or Daily Homework Quiz: TE pages 503 and
        490, CRB page 11, or Transparencies

TEACHING OPTIONS

____ Motivating the Lesson: TE page 504
____ Lesson Opener (Application): CRB page 12 or Transparencies
____ Graphing Calculator Activity with Keystrokes: CRB page 13
____ Examples 1–7: SE pages 503–506
____ Extra Examples: TE pages 504–506 or Transparencies
____ Closure Question: TE page 506
____ Guided Practice Exercises: SE page 507

APPLY/HOMEWORK

Homework Assignment

____ Block Schedule: 23, 26, 29, and every 3rd Ex. through 77, 79–81, 87, 95, 96, 104, 107, 110, 111

Reteaching the Lesson

____ Practice Masters: CRB pages 14–16 (Level A, Level B, Level C)
____ Reteaching with Practice: CRB pages 17–18 or Practice Workbook with Examples
____ Personal Student Tutor

Extending the Lesson

____ Applications (Interdisciplinary): CRB page 20
____ Challenge: SE page 510; CRB page 21 or Internet

ASSESSMENT OPTIONS

____ Checkpoint Exercises: TE pages 504–506 or Transparencies
____ Daily Homework Quiz (9.1): TE page 510, CRB page 24, or Transparencies
____ Standardized Test Practice: SE page 509; TE page 510; STP Workbook; Transparencies

Notes _____

_____

_____

CHAPTER PACING GUIDE	
Day	Lesson
1	Assess Ch. 8; **9.1 (all)**
2	9.2 (all); 9.3 (begin)
3	9.3 (end); 9.4 (all)
4	9.5 (all)
5	9.6 (all)
6	9.7 (all); 9.8 (begin)
7	9.8 (end); Review Ch. 9
8	Assess Ch. 9; 10.1 (all)

Algebra 1
Chapter 9 Resource Book

NAME _____ DATE _____

WARM-UP EXERCISES

For use before Lesson 9.1, pages 503–510

Simplify.

1. 6^2

2. $(-14)^2$

3. -9^2

4. 0^2

5. $-4x^2$, for $x = 3$

DAILY HOMEWORK QUIZ

For use after Lesson 8.6, pages 483–492

1. In a factory, a piece of machinery that cost $45,000 depreciates at the rate of 16% per year. Find the value of the machinery 5 years from the date of purchase.

Classify the model as *exponential growth* or *exponential decay*. Identify the growth or decay factor and the percent of increase or decrease per time period.

2. $y = 120(0.89)^t$

3. $y = 93(1.03)^t$

4. The concentration y of a certain medication in a person's bloodstream t hours after taking A mg of the medication is modeled by $y = A(0.75)^t$. Find the amount of the medication remaining in a person's bloodstream 3 h after taking 1000 mg of the medication.

NAME _____ DATE _____

Application Lesson Opener

For use with pages 503–510

Find the area of the square.

1.

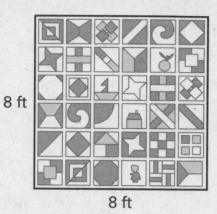

8 ft

8 ft

2.

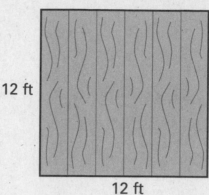

12 ft

12 ft

3. Suppose you knew the area of each square in Questions 1 and 2 and were asked to find the length of a side. Explain how you would do this.

Find the length of a side of the square whose area is given.

4. Area = 49 in.^{2}

5. Area = 25 in.^{2}

6. Area = 36 ft^2

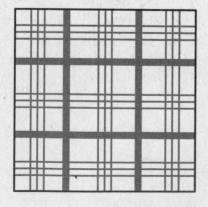

7. Area = 9 in.^{2}

Graphing Calculator Activity Keystrokes

For use with page 504.

Keystrokes for Example 4

TI-82

(1 + 2 × 2nd [√] 3)

÷ 4 ENTER

(1 − 2 × 2nd [√] 3)

÷ 4 ENTER

TI-83

(1 + 2 × 2nd [√] 3)

÷ 4 ENTER

(1 − 2 × 2nd [√] 3)

÷ 4 ENTER

SHARP EL-9600c

(1 + 2 × 2ndF [√] 3 ▶

) ÷ 4

ENTER

(1 − 2 × 2ndF [√] 3 ▶

) ÷ 4

ENTER

CASIO CFX-9850GA PLUS

From the main menu, choose RUN.

(1 + 2 × SHIFT [√] 3)

÷ 4 EXE

(1 − 2 × SHIFT [√] 3)

÷ 4 EXE

NAME _____ DATE _____

Practice A

For use with pages 503–510

Find all square roots of the number or write *no square roots*.

1. 64   **2.** -16 **3.** 121   **4.** 1

5. -25 **6.** 0.16   **7.** 49   **8.** 0

Evaluate the expression. Give the exact value if possible. Otherwise, approximate to the nearest hundredth.

9. $\sqrt{16}$ **10.** $-\sqrt{64}$ **11.** $-\sqrt{49}$ **12.** $-\sqrt{225}$

13. $-\sqrt{36}$ **14.** $\sqrt{144}$ **15.** $\sqrt{32}$ **16.** $-\sqrt{1.69}$

Evaluate $\sqrt{b^2 - 4ac}$ for the given values.

17. $a = 1, b = 5, c = -6$ **18.** $a = 2, b = -9, c = 7$ **19.** $a = 4, b = -3, c = -1$

20. $a = 10, b = -21, c = 9$ **21.** $a = -5, b = 8, c = 2$ **22.** $a = 3, b = -4, c = 5$

Use a calculator to evaluate the expression. Round the results to the nearest hundredth.

23. $\dfrac{6 \pm 2\sqrt{3}}{5}$ **24.** $\dfrac{3 \pm 5\sqrt{2}}{4}$ **25.** $\dfrac{-5 \pm 2\sqrt{3}}{3}$

26. $\dfrac{5 \pm 3\sqrt{3}}{2}$ **27.** $\dfrac{8 \pm 3\sqrt{2}}{-2}$ **28.** $\dfrac{-2 \pm 3\sqrt{5}}{7}$

Solve the equation or write *no solution*. Write the solutions as integers if possible. Otherwise, write them as radical expressions.

29. $x^2 = 25$ **30.** $y^2 = 81$ **31.** $3a^2 = 147$

32. $-3x^2 = 27$ **33.** $x^2 + 4 = 16$ **34.** $2b^2 - 7 = -7$

Falling Object Model   **In Exercises 35–38, an object is dropped from a height *s*. How long does it take to reach the ground? Assume there is no air resistance.**

35. $s = 64$ feet   **36.** $s = 576$ feet   **37.** $s = 784$ feet

Algebra 1
Chapter 9 Resource Book

NAME _____ DATE _____

Practice B
For use with pages 503–510

Evaluate the expression. Give the exact value if possible. Otherwise, approximate to the nearest hundredth.

1. $-\sqrt{49}$ **2.** $-\sqrt{81}$ **3.** $\sqrt{0.25}$ **4.** $-\sqrt{38}$

5. $\pm\sqrt{0.16}$ **6.** $\sqrt{12.25}$ **7.** $\pm\sqrt{0.4}$ **8.** $\sqrt{19.36}$

Evaluate $\sqrt{b^2 - 4ac}$ for the given values.

9. $a = 3, b = -6, c = 3$ **10.** $a = 5, b = 8, c = 3$ **11.** $a = -4, b = -9, c = -6$

12. $a = 4.25, b = -10, c = 5$ **13.** $a = -6, b = 5, c = -4$ **14.** $a = -2, b = 9, c = 5$

Use a calculator to evaluate the expression. Round the results to the nearest hundredth.

15. $\dfrac{8 \pm 3\sqrt{6}}{-1}$ **16.** $\dfrac{9 \pm 2\sqrt{12}}{6}$ **17.** $\dfrac{6 \pm 4\sqrt{5}}{7}$

18. $\dfrac{6 \pm 3\sqrt{2}}{3}$ **19.** $\dfrac{9 \pm 5\sqrt{8}}{-2}$ **20.** $\dfrac{-3 \pm 0.2\sqrt{7}}{4}$

Solve the equation or write *no solution*. Write the solutions as integers if possible. Otherwise, write them as radical expressions.

21. $x^2 = 49$ **22.** $3y^2 = 192$ **23.** $7a^2 = 0$

24. $6 - 3x^2 = 27$ **25.** $y^2 + 19 = 33$ **26.** $8b^2 - 16 = 24$

27. $3x^2 + 9 = 84$ **28.** $2x^2 - 7 = 1$ **29.** $-4x^2 + 6 = -394$

30. *Geometry*   Use the volume V to find the length of the radius. (Use $\pi \approx 3.14$.)

Cylinder: $r = \sqrt{\dfrac{V}{\pi h}}$

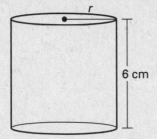

6 cm

Volume $= 230.79 \text{ cm}^3$

31. *Falling Object*   An object is dropped from a height of 576 feet. How long does it take for the object to reach the ground? Assume there is no air resistance.

**Evaluate the expression. Give the exact value if possible.
Otherwise, approximate to the nearest hundredth.**

1. $\sqrt{144}$ **2.** $-\sqrt{0.5}$ **3.** $\sqrt{3.25}$ **4.** $-\sqrt{54}$

5. $\pm\sqrt{0.36}$ **6.** $\sqrt{36.5}$ **7.** $\pm\sqrt{0.9}$ **8.** $\sqrt{73.96}$

Evaluate $\sqrt{b^2 - 4ac}$ for the given values.

9. $a = -15, b = -8, c = -1$ **10.** $a = 6, b = 9, c = -4$ **11.** $a = 4, b = -6, c = 2$

12. $a = 15, b = -10, c = 1$ **13.** $a = 5, b = -12, c = 10$ **14.** $a = -2.75, b = 8, c = -5$

**Use a calculator to evaluate the expression. Round the results to
the nearest hundredth.**

15. $\dfrac{11 \pm 5\sqrt{7}}{-1}$ **16.** $\dfrac{10 \pm 9\sqrt{12}}{3}$ **17.** $\dfrac{-6 \pm 2\sqrt{19}}{12}$

18. $\dfrac{4 \pm 7\sqrt{2}}{-16}$ **19.** $\dfrac{20 \pm 15\sqrt{8}}{-5}$ **20.** $\dfrac{-3 \pm 1.5\sqrt{10}}{4}$

**Solve the equation or write *no solution*. Write the solutions as
integers if possible. Otherwise, write them as radical expressions.**

21. $x^2 = 100$ **22.** $2y^2 = 32$ **23.** $\frac{2}{3}a^2 = 6$

24. $10 - 2x^2 = 3$ **25.** $7y^2 + 14 = 0$ **26.** $3b^2 - 6 = 9$

27. $\frac{1}{2}x^2 - 7 = 1$ **28.** $2x^2 + 5 = 9$ **29.** $\frac{1}{7}x^2 - 2 = 5$

**In Exercises 30–35, use a calculator to solve the equation or write
no solution. Round the results to the nearest hundredth.**

30. $6x^2 = 48$ **31.** $2x^2 + 4 = 96$ **32.** $\frac{1}{2}x^2 - 12 = 21$

33. $-2x^2 + 30 = 18$ **34.** $5x^2 - 15 = 45$ **35.** $\frac{2}{3}x^2 + 25 = 87$

36. *Falling on Mars*   The height h (in feet) of a falling object on Mars after
t seconds is given by $h = -6t^2 + s$, where s is the initial height from
which the object was dropped. A rock is dropped from a space probe
from a height of 90 feet. About how long would it take to reach Mars's
surface? If the same rock is dropped on Earth, about how long would it
take to reach the Earth's surface? Assume there is no air resistance.

37. *Falling Object*   An object is dropped from a height of 288 feet. How
long does it take for the object to reach the ground? Assume there is no
air resistance.

Reteaching with Practice

For use with pages 503–510

GOAL Evaluate and approximate square roots and solve a quadratic equation by finding square roots

---

VOCABULARY

If $b^2 = a$, then b is a **square root** of a.

A square root b can be a **positive square root** (or a principal square root) or a **negative square root**.

A **radicand** is a number or expression inside a radical symbol $\sqrt{}$.

Perfect squares are numbers whose square roots are integers or quotients of integers.

An **irrational number** is a number that cannot be written as the quotient of two integers.

A **radical expression** involves square roots (or *radicals*).

A **quadratic equation** is an equation that can be written in the **standard form** $ax^2 + bx + c = 0$, where $a \neq 0$. In standard form, a is the **leading coefficient**.

EXAMPLE 1 *Finding Square Roots of Numbers*

Evaluate the expression.

a. $\sqrt{81}$ **b.** $-\sqrt{49}$ **c.** $\pm\sqrt{0.16}$ **d.** $\sqrt{-1}$

SOLUTION

a. $\sqrt{81} = 9$ Positive square root

b. $-\sqrt{49} = -7$ Negative square root

c. $\pm\sqrt{0.16} = \pm 0.4$ Two square roots

d. $\sqrt{-1}$ (undefined)   No real square root

Exercises for Example 1

Evaluate the expression.

1. $\sqrt{0.09}$ **2.** $\sqrt{36}$ **3.** $-\sqrt{25}$ **4.** $\pm\sqrt{100}$

NAME _____ DATE _____

Reteaching with Practice

For use with pages 503–510

EXAMPLE 2 **Evaluating a Radical Expression**

Evaluate $\sqrt{b^2 - 4ac}$ when $a = -2$, $b = -5$, and $c = 2$.

SOLUTION

$$\sqrt{b^2 - 4ac} = \sqrt{(-5)^2 - 4(-2)(2)} \qquad \text{Substitute values.}$$

$$= \sqrt{25 + 16} \qquad\qquad \text{Simplify.}$$

$$= \sqrt{41} \qquad\qquad \text{Simplify.}$$

$$\approx 6.40 \qquad\qquad \text{Round to the nearest hundredth.}$$

Exercises for Example 2

Evaluate $\sqrt{b^2 - 4ac}$ for the given values.

5. $a = -3$, $b = 6$, $c = -3$ **6.** $a = 1$, $b = 5$, $c = 4$

EXAMPLE 3 **Rewriting Before Finding Square Roots**

Solve $4x^2 - 100 = 0$.

SOLUTION

$$4x^2 - 100 = 0 \qquad \text{Write original equation.}$$

$$4x^2 = 100 \qquad \text{Add 100 to each side.}$$

$$x^2 = 25 \qquad \text{Divide each side by 4.}$$

$$x = \pm\sqrt{25} \qquad \text{Find square roots.}$$

$$x = \pm 5 \qquad \text{25 is a perfect square.}$$

Exercises for Example 3

Solve the equation or write *no solution*. Write the solutions as integers if possible. Otherwise write them as radical expressions.

7. $6x^2 - 54 = 0$ **8.** $5x^2 - 15 = 0$ **9.** $2x^2 - 98 = 0$

NAME _____ DATE _____

Quick Catch-Up for Absent Students

For use with pages 503–510

The items checked below were covered in class on (date missed) _____

Lesson 9.1: Solving Quadratic Equations by Finding Square Roots

____ **Goal 1:** Evaluate and approximate square roots. (pp. 503–504)

Material Covered:

____ Example 1: Finding Square Roots of Numbers

____ Student Help: Square Root Table

____ Example 2: Evaluating Square Roots

____ Example 3: Evaluating a Radical Expression

____ Student Help: Keystroke Help

____ Example 4: Evaluating an Expression with a Calculator

Vocabulary:

square root, p. 503   positive square root, p. 503

negative square root, p. 503   radicand, p. 503

perfect square, p. 504   irrational number, p. 504

radical expression, p. 504

____ **Goal 2:** Solve a quadratic equation by finding square roots. (pp. 505–506)

Material Covered:

____ Example 5: Solving Quadratic Equations

____ Student Help: Study Tip

____ Example 6: Rewriting Before Finding Square Roots

____ Student Help: Look Back

____ Example 7: Using a Falling Object Model

Vocabulary:

quadratic equation, p. 505   leading coefficient, p. 505

standard form of a quadratic equation, p. 505

____ Other (specify) _____

_____

Homework and Additional Learning Support

____ Textbook (specify) pp. 507–510 _____

_____

____ *Reteaching with Practice* worksheet (specify exercises)_____

____ *Personal Student Tutor* for Lesson 9.1

Interdisciplinary Application

For use with pages 503–510

Right Circular Cylinder

GEOMETRY   A cylinder, in geometry, is a solid figure with two identical bases that lie on parallel planes. Each base is bounded by a curved edge. The lateral surface (side) of a cylinder consists of parallel lines that join corresponding points on each base. When the two bases of a cylinder are circles, the figure is called a circular cylinder. A right circular cylinder is a circular cylinder with a lateral surface that is perpendicular to the bases.

The height h of a cylinder is the perpendicular distance between the planes containing the bases. The volume V of a cylinder can be calculated by multiplying the height by the area enclosed by either of the two bases. If the bases are circles, then the area of a base is equal to pi, π, times the square of the radius, r. The formula for the volume of a right circular cylinder can be written as $V = \pi \cdot r^2 \cdot h$. An approximate value of pi, π, is 3.1416.

1. Rewrite the formula $V = \pi r^2 h$, solving for r.

2. Using your new formula in Exercise 1, find the radius of a right circular cylinder with a height of 8 centimeters and a volume of 904.7808 cubic centimeters.

3. A right circular cylinder has a volume of 33,866.448 cubic feet and a height of 55 feet. Find the radius of the cylinder.

4. A right circular cylinder is twice as tall as another cylinder, but only half as wide. Which cylinder has a greater volume?

5. A chemical company has developed an additive that cuts down on friction, permitting water to flow through a hose twice as fast. Using this additive, a fire department can replace 7-centimeter hoses with 5-centimeter hoses. This will make the hose considerably lighter and still deliver the same amount of water. Show that the amount of water in a hose with a 7-centimeter diameter is about twice the amount of water in a hose with a 5-centimeter diameter. (Use the same length for each hose.)

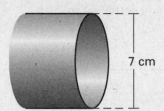
7 cm

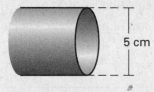
5 cm

NAME _____ DATE _____

Challenge: Skills and Applications

For use with pages 503–510

In Exercises 1–8, solve the equation. Write the solution as integers if possible. Otherwise write them as radical expressions.

Example: $(x + 3)^2 = 5$

Solution: $(x + 3)^2 = 5 \Rightarrow x + 3 = \pm\sqrt{5} \Rightarrow x = -3 \pm \sqrt{5}$

1. $(x - 2)^2 = 25$　　　　　　　**2.** $(x + 5)^2 = 9$

3. $(x - 7)^2 = 10$　　　　　　　**4.** $3(x + 1)^2 = 24$

5. $10 - 2(x - 1)^2 = 4$　　　　**6.** $\frac{1}{2}(x - 4)^2 + 3 = 11$

7. $-3(x + 5)^2 + 17 = 2$　　　　**8.** $5\left(x + \frac{3}{4}\right)^2 - 2 = 43$

In Exercises 9–11, use the following information.

Around 1590, the Italian scientist Galileo is said to have simultaneously dropped two balls of different weights from the Leaning Tower of Pisa and found that they reached the ground at the same time. If the balls were dropped from a height of 20 meters, their height h above the ground t seconds after their release is given by:

$$h = -4.9t^2 + 20$$

9. How long would it take the balls to reach the ground?

10. Rewrite the equation for dropping the balls from a height of 17 meters.

11. Suppose Galileo released a third ball from the same height 0.5 seconds after the first two balls. Write an expression for the height of this ball above the ground t seconds *after the release of the first two balls.*

TEACHER'S NAME _____ CLASS _____ ROOM _____ DATE _____

Lesson Plan

1-day lesson (See *Pacing the Chapter*, TE pages 500C–500D)   **For use with pages 511–516**

 GOALS   1. **Use properties of radicals to simplify radicals.**
2. **Use quadratic equations to model real-life problems.**

State/Local Objectives _____

_____

✓ **Check the items you wish to use for this lesson.**

STARTING OPTIONS
____ Homework Check: TE page 507; Answer Transparencies
____ Warm-Up or Daily Homework Quiz: TE pages 511 and 510, CRB page 24, or Transparencies

TEACHING OPTIONS
____ Motivating the Lesson: TE page 512
____ Lesson Opener (Activity): CRB page 25 or Transparencies
____ Examples 1–3: SE pages 512–513
____ Extra Examples: TE pages 512–513 or Transparencies; Internet
____ Closure Question: TE page 513
____ Guided Practice Exercises: SE page 514

APPLY/HOMEWORK
Homework Assignment
____ Basic   10–52 even, 58–60, 65, 70, 74, 75
____ Average   10–52 even, 53, 54, 58–60, 65, 70, 74, 75
____ Advanced   10–52 even, 55–65, 70, 74, 75

Reteaching the Lesson
____ Practice Masters: CRB pages 26–28 (Level A, Level B, Level C)
____ Reteaching with Practice: CRB pages 29–30 or Practice Workbook with Examples
____ Personal Student Tutor

Extending the Lesson
____ Applications (Real-Life): CRB page 32
____ Challenge: SE page 516; CRB page 33 or Internet

ASSESSMENT OPTIONS
____ Checkpoint Exercises: TE pages 512–513 or Transparencies
____ Daily Homework Quiz (9.2): TE page 516, CRB page 36, or Transparencies
____ Standardized Test Practice: SE page 516; TE page 516; STP Workbook; Transparencies

Notes _____

_____

_____

TEACHER'S NAME _____ CLASS _____ ROOM _____ DATE _____

Lesson Plan for Block Scheduling

Half-day lesson (See *Pacing the Chapter,* TE pages 500C–500D)   For use with pages 511–516

GOALS
1. **Use properties of radicals to simplify radicals.**
2. **Use quadratic equations to model real-life problems.**

State/Local Objectives _____

_____

_____

✓ **Check the items you wish to use for this lesson.**

STARTING OPTIONS

____ Homework Check: TE page 507; Answer Transparencies
____ Warm-Up or Daily Homework Quiz: TE pages 511 and
        510, CRB page 24, or Transparencies

TEACHING OPTIONS

____ Motivating the Lesson: TE page 512
____ Lesson Opener (Activity): CRB page 25 or Transparencies
____ Examples 1–3: SE pages 512–513
____ Extra Examples: TE pages 512–513 or Transparencies; Internet
____ Closure Question: TE page 513
____ Guided Practice Exercises: SE page 514

APPLY/HOMEWORK
Homework Assignment (See also the assignment for Lesson 9.3.)
____ Block Schedule: 10–52 even, 53, 54, 58–60, 65, 70, 74, 75

Reteaching the Lesson
____ Practice Masters: CRB pages 26–28 (Level A, Level B, Level C)
____ Reteaching with Practice: CRB pages 29–30 or Practice Workbook with Examples
____ Personal Student Tutor

Extending the Lesson
____ Applications (Real-Life): CRB page 32
____ Challenge: SE page 516; CRB page 33 or Internet

ASSESSMENT OPTIONS

____ Checkpoint Exercises: TE pages 512–513 or Transparencies
____ Daily Homework Quiz (9.2): TE page 516, CRB page 36, or Transparencies
____ Standardized Test Practice: SE page 516; TE page 516; STP Workbook; Transparencies

Notes _____

_____

_____

CHAPTER PACING GUIDE

Day	Lesson
1	Assess Ch. 8; 9.1 (all)
2	**9.2 (all)**; 9.3 (begin)
3	9.3 (end); 9.4 (all)
4	9.5 (all)
5	9.6 (all)
6	9.7 (all); 9.8 (begin)
7	9.8 (end); Review Ch. 9
8	Assess Ch. 9; 10.1 (all)

Lesson 9.2

NAME _____ DATE _____

WARM-UP EXERCISES

For use before Lesson 9.2, pages 511–516

Lesson 9.2

Evaluate each expression.

1. $\sqrt{25}$

2. $-\sqrt{100}$

3. $\sqrt{16} \cdot \sqrt{4}$

4. $\dfrac{2}{3} \cdot \sqrt{9}$

5. $\dfrac{\sqrt{100}}{5}$

DAILY HOMEWORK QUIZ

For use after Lesson 9.1, pages 503–510

1. Find all squares roots of 0.25.

2. Evaluate $-\sqrt{1.44}$.

3. Evaluate $\sqrt{b^2 - 4ac}$ for $a = 3$, $b = 3$, and $c = -6$.

4. Using a calculator, evaluate $\dfrac{-3 \pm 2\sqrt{5}}{4}$ to the nearest hundredth.

5. Solve $x^2 - 7 = 13$. Write the result as a radical expression.

6. Solve $3y^2 + 4 = 11$. Write the result to the nearest hundredth.

7. A cliff diver jumps from an 84 ft cliff. Write a falling object model for the time t it will take the diver to hit the water. Assume there is no air resistance.

NAME _____ DATE _____

Activity Lesson Opener

For use with pages 511–516

SET UP: Work with a partner.

YOU WILL NEED: • markers

1. Identify the card that has been assigned to your group.

Card 1

$\sqrt{4 \cdot 5}$	$\sqrt{\dfrac{3}{4}}$	$\sqrt{8} \cdot \sqrt{6}$
$\sqrt{\dfrac{7}{6}}$	$\dfrac{\sqrt{2}}{\sqrt{5}}$	$\sqrt{5 \cdot 8}$
$\sqrt{5} \cdot \sqrt{7}$	$\dfrac{\sqrt{10}}{\sqrt{7}}$	$\sqrt{3 \cdot 2}$

Card 2

$\dfrac{\sqrt{10}}{\sqrt{7}}$	$\sqrt{3 \cdot 2}$	$\sqrt{5 \cdot 8}$
$\sqrt{4 \cdot 5}$	$\sqrt{\dfrac{7}{6}}$	$\sqrt{5} \cdot \sqrt{7}$
$\sqrt{8} \cdot \sqrt{6}$	$\sqrt{\dfrac{3}{4}}$	$\dfrac{\sqrt{2}}{\sqrt{5}}$

Card 3

$\sqrt{5} \cdot \sqrt{7}$	$\sqrt{5 \cdot 8}$	$\sqrt{\dfrac{7}{6}}$
$\dfrac{\sqrt{2}}{\sqrt{5}}$	$\sqrt{\dfrac{3}{4}}$	$\sqrt{8} \cdot \sqrt{6}$
$\sqrt{3 \cdot 2}$	$\dfrac{\sqrt{10}}{\sqrt{7}}$	$\sqrt{4 \cdot 5}$

Card 4

$\sqrt{\dfrac{3}{4}}$	$\sqrt{5} \cdot \sqrt{7}$	$\sqrt{3 \cdot 2}$
$\sqrt{5 \cdot 8}$	$\sqrt{8} \cdot \sqrt{6}$	$\sqrt{4 \cdot 5}$
$\dfrac{\sqrt{10}}{\sqrt{7}}$	$\sqrt{\dfrac{7}{6}}$	$\dfrac{\sqrt{2}}{\sqrt{5}}$

2. As your teacher writes one of the expressions on the overhead or chalkboard, find the expression on your card with the same value and place a marker on that square. The first group to cover three squares in a row, column, or diagonal wins!

$\dfrac{\sqrt{7}}{\sqrt{6}}$	$\sqrt{5} \cdot \sqrt{8}$	$\sqrt{\dfrac{10}{7}}$
$\sqrt{4} \cdot \sqrt{5}$	$\dfrac{\sqrt{3}}{\sqrt{4}}$	$\sqrt{3} \cdot \sqrt{2}$
$\sqrt{5 \cdot 7}$	$\sqrt{\dfrac{2}{5}}$	$\sqrt{8 \cdot 6}$

Practice A

For use with pages 511–516

Match the radical expression with its simplified form.

A. $6\sqrt{2}$ B. $4\sqrt{5}$ C. $4\sqrt{2}$ D. $5\sqrt{3}$

E. $2\sqrt{7}$ F. $8\sqrt{2}$ G. $3\sqrt{3}$ H. $7\sqrt{2}$

1. $\sqrt{128}$ **2.** $\sqrt{80}$ **3.** $\sqrt{98}$ **4.** $\sqrt{72}$

5. $\sqrt{27}$ **6.** $\sqrt{32}$ **7.** $\sqrt{75}$ **8.** $\sqrt{28}$

Use the product property to simplify the expression.

9. $\sqrt{20}$ **10.** $\sqrt{12}$ **11.** $\sqrt{40}$ **12.** $\sqrt{18}$

13. $\sqrt{48}$ **14.** $\sqrt{54}$ **15.** $\sqrt{45}$ **16.** $\sqrt{3} \cdot \sqrt{4}$

Use the quotient property to simplify the expression.

17. $\sqrt{\dfrac{3}{4}}$ **18.** $\sqrt{\dfrac{9}{16}}$ **19.** $\sqrt{\dfrac{3}{25}}$ **20.** $\sqrt{\dfrac{20}{49}}$

21. $3\sqrt{\dfrac{6}{3}}$ **22.** $2\sqrt{\dfrac{7}{9}}$ **23.** $5\sqrt{\dfrac{13}{25}}$ **24.** $10\sqrt{\dfrac{11}{36}}$

Simplify the expression.

25. $\dfrac{\sqrt{64}}{\sqrt{36}}$ **26.** $\dfrac{\sqrt{80}}{\sqrt{4}}$ **27.** $\dfrac{\sqrt{18}}{\sqrt{81}}$ **28.** $\dfrac{\sqrt{75}}{\sqrt{25}}$

29. $\dfrac{\sqrt{40}}{\sqrt{16}}$ **30.** $\dfrac{\sqrt{100}}{\sqrt{25}}$ **31.** $\sqrt{20} \cdot \sqrt{80}$ **32.** $3\sqrt{12} \cdot \sqrt{3}$

Geometry   **Find the area of the figure. Give both the exact answer in simplified form and the decimal approximation rounded to the nearest hundredth.**

33.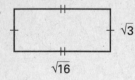

34.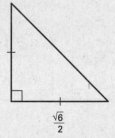

35.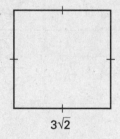

36. *Boat Racing*   The maximum speed s (in knots or nautical miles per hour) that some kinds of boats can travel can be modeled by $s^2 = \frac{16}{9}x$, where x is the length of the water line in feet. Find the maximum speed of a sailboat with a 30-foot water line.

NAME _____ DATE _____

Practice B

For use with pages 511–516

Match the radical expression with its simplified form.

A. $5\sqrt{2}$ **B.** $4\sqrt{5}$ **C.** $3\sqrt{2}$ **D.** $7\sqrt{3}$

E. $4\sqrt{7}$ **F.** $6\sqrt{2}$ **G.** $6\sqrt{3}$ **H.** $5\sqrt{5}$

1. $\sqrt{108}$ **2.** $\sqrt{72}$ **3.** $\sqrt{147}$ **4.** $\sqrt{80}$

5. $\sqrt{18}$ **6.** $\sqrt{112}$ **7.** $\sqrt{125}$ **8.** $\sqrt{50}$

Use the product property to simplify the expression.

9. $\sqrt{50}$ **10.** $\sqrt{20}$ **11.** $\sqrt{240}$ **12.** $\sqrt{108}$

13. $\sqrt{300}$ **14.** $\sqrt{3} \cdot \sqrt{12}$ **15.** $\dfrac{1}{3}\sqrt{45}$ **16.** $\dfrac{1}{2}\sqrt{128}$

Use the quotient property to simplify the expression.

17. $\sqrt{\dfrac{16}{25}}$ **18.** $\sqrt{\dfrac{1}{9}}$ **19.** $7\sqrt{\dfrac{3}{16}}$ **20.** $5\sqrt{\dfrac{20}{49}}$

21. $2\sqrt{\dfrac{15}{3}}$ **22.** $\dfrac{\sqrt{40}}{14}$ **23.** $10\sqrt{\dfrac{20}{64}}$ **24.** $\sqrt{\dfrac{9}{81}}$

Simplify the expression.

25. $\dfrac{\sqrt{28}}{\sqrt{49}}$ **26.** $\sqrt{\dfrac{12}{16}}$ **27.** $\dfrac{\sqrt{64}}{\sqrt{4}}$ **28.** $\dfrac{\sqrt{9}}{\sqrt{81}}$

29. $\dfrac{\sqrt{56}}{\sqrt{36}}$ **30.** $\dfrac{\sqrt{112}}{\sqrt{100}}$ **31.** $2\sqrt{42} \cdot \sqrt{9}$ **32.** $\dfrac{1}{3}\sqrt{18} \cdot \sqrt{3}$

33. $\sqrt{3} \cdot \dfrac{\sqrt{20}}{\sqrt{5}}$ **34.** $8\sqrt{27} \cdot \sqrt{72}$ **35.** $\dfrac{\sqrt{6} \cdot \sqrt{36}}{\sqrt{2}}$ **36.** $\dfrac{-4 \cdot \sqrt{45}}{\sqrt{144}}$

Geometry **Find the area of the figure. Give both the exact answer in simplified form and the decimal approximation rounded to the nearest hundredth.**

37.
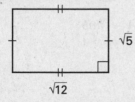
$\sqrt{5}$
$\sqrt{12}$

38.
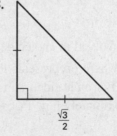
$\dfrac{\sqrt{3}}{2}$

39.
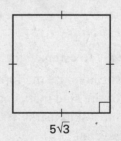
$5\sqrt{3}$

Use the product property to simplify the expression.

1. $\sqrt{80}$

2. $\sqrt{245}$

3. $\sqrt{112}$

4. $\sqrt{288}$

5. $\sqrt{27} \cdot \sqrt{18}$

6. $\sqrt{25} \cdot \sqrt{28}$

7. $\frac{1}{4}\sqrt{80}$

8. $\frac{2}{3}\sqrt{98}$

Use the quotient property to simplify the expression.

9. $\sqrt{\dfrac{25}{100}}$

10. $\sqrt{\dfrac{1}{16}}$

11. $10\sqrt{\dfrac{5}{36}}$

12. $21\sqrt{\dfrac{48}{49}}$

13. $3\sqrt{\dfrac{21}{7}}$

14. $\dfrac{\sqrt{90}}{33}$

15. $5\sqrt{\dfrac{32}{121}}$

16. $\sqrt{\dfrac{16}{169}}$

Simplify the expression.

17. $\dfrac{\sqrt{72}}{\sqrt{64}}$

18. $\sqrt{\dfrac{8}{25}}$

19. $\dfrac{\sqrt{144}}{\sqrt{9}}$

20. $\dfrac{\sqrt{25}}{\sqrt{100}}$

21. $\dfrac{\sqrt{63}}{\sqrt{49}}$

22. $\sqrt{\dfrac{6}{121}}$

23. $5\sqrt{105} \cdot \sqrt{16}$

24. $\dfrac{1}{5}\sqrt{50} \cdot \sqrt{2}$

25. $\sqrt{17} \cdot \dfrac{\sqrt{112}}{\sqrt{7}}$

26. $8\sqrt{153} \cdot \sqrt{196}$

27. $\dfrac{\sqrt{81} \cdot \sqrt{10}}{\sqrt{5}}$

28. $\dfrac{-7\sqrt{63}}{\sqrt{256}}$

Geometry   **Find the area of the figure. Give both the exact answer in simplified form and the decimal approximation rounded to the nearest hundredth. For approximations, use $\pi \approx 3.14$.**

29.

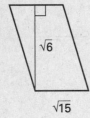

$A = bh$

30.

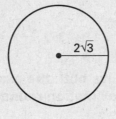

$A = \pi r^2$

31.

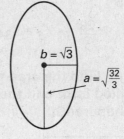

$A = \pi ab$

32. *Speed*   To estimate the speed s of a car involved in an accident, investigators use the formula $s = \frac{11}{2}\sqrt{\frac{3}{4}l}$, where l represents the length in feet of tire skid marks on the pavement. After an accident, an investigator measures skid marks 160 feet long. Approximately how fast was the car traveling?

Reteaching with Practice

For use with pages 511–516

GOAL **Use properties of radicals to simplify radicals and use quadratic equations to model real-life problems**

VOCABULARY

Product Property   The square root of a product equals the product of the square roots of the factors.

$$\sqrt{ab} = \sqrt{a} \cdot \sqrt{b} \text{ when } a \text{ and } b \text{ are positive numbers}$$

Quotient Property   The square root of a quotient equals the quotient of the square roots of the numerator and denominator.

$$\sqrt{\frac{a}{b}} = \frac{\sqrt{a}}{\sqrt{b}} \text{ when } a \text{ and } b \text{ are positive numbers}$$

An expression with radicals is in **simplest form** if the following are true:

• No perfect square factors other than 1 are in the radicand.

• No fractions are in the radicand.

• No radicals appear in the denominator of a fraction.

EXAMPLE 1   *Simplifying with the Product Property*

Simplify the expression $\sqrt{147}$.

SOLUTION

You can use the product property to simplify a radical by removing perfect square factors from the radicand.

$$
\begin{aligned}
\sqrt{147} &= \sqrt{49 \cdot 3} && \text{Factor using perfect square factor.} \\
&= \sqrt{49} \cdot \sqrt{3} && \text{Use product property.} \\
&= 7\sqrt{3} && \text{Simplify.}
\end{aligned}
$$

Exercises for Example 1
....
Simplify the expression.

1. $\sqrt{98}$ **2.** $\sqrt{52}$ **3.** $\sqrt{300}$ **4.** $\sqrt{99}$

Reteaching with Practice

For use with pages 511–516

EXAMPLE 2 *Simplifying with the Quotient Property*

Simplify the expression $\dfrac{\sqrt{63}}{6}$.

SOLUTION

$$\dfrac{\sqrt{63}}{6} = \dfrac{\sqrt{9 \cdot 7}}{6} \qquad \text{Factor using perfect square factor.}$$

$$= \dfrac{3\sqrt{7}}{6} \qquad \text{Remove perfect square factor.}$$

$$= \dfrac{\sqrt{7}}{2} \qquad \text{Divide out common factors.}$$

Exercises for Example 2

Simplify the expression.

5. $\sqrt{\dfrac{11}{4}}$ **6.** $\dfrac{\sqrt{200}}{60}$ **7.** $\sqrt{\dfrac{5}{9}}$ **8.** $\dfrac{\sqrt{75}}{20}$

EXAMPLE 3 *Simplifying Radical Expressions*

The speed s (in meters per second) at which a tsunami moves is determined by the depth d (in meters) of the ocean: $s = \sqrt{gd}$, where g is 9.8 $m/\sec^2$. Find the speed of a tsunami in a region of the ocean that is 2000 meters deep. Write the result in simplified form.

SOLUTION

Write the model for speed of the tsunami and let $d = 2000$ meters.

$$s = \sqrt{gd} \qquad \text{Write model.}$$

$$= \sqrt{(9.8)(2000)} \qquad \text{Substitute 9.8 for } g \text{ and 2000 for } d.$$

$$= \sqrt{19{,}600} \qquad \text{Simplify.}$$

$$= \sqrt{196 \cdot 100} \qquad \text{Factor using perfect square factors.}$$

$$= 14 \cdot 10 \qquad \text{Find square roots.}$$

$$= 140 \qquad \text{Simplify.}$$

The speed of the tsunami is 140 meters per second.

Exercise for Example 3

9. Rework Example 3 to find the speed of a tsunami in a region of the ocean that is 500 meters deep. Write the result in simplified form.

NAME _____ DATE _____

Quick Catch-Up for Absent Students

For use with pages 511–516

The items checked below were covered in class on (date missed) _____

Lesson 9.2: Simplifying Radicals

_____ **Goal 1:** Use properties of radicals to simplify radicals. (pp. 511–512)

Material Covered:

    _____ Activity: Investigating Properties of Radicals

    _____ Student Help: Skills Review

    _____ Example 1: Simplifying with the Product Property

    _____ Example 2: Simplifying with the Quotient Property

Vocabulary:

    simplest form of a radical expression, p. 512

_____ **Goal 2:** Use quadratic equations to model real-life problems. (p. 513)

Material Covered:

    _____ Example 3: Simplifying Radical Expressions

_____ Other (specify) _____

_____

Homework and Additional Learning Support

    _____ Textbook (specify) <u>pp. 514–516</u> _____

    _____

    _____ Internet: Extra Examples at www.mcdougallittel.com

    _____ *Reteaching with Practice* worksheet (specify exercises)_____

    _____ *Personal Student Tutor* for Lesson 9.2

Real-Life Application:
When Will I Ever Use This?

For use with pages 511–516

Centripetal Acceleration

A car traveling with a constant speed on a circular track is accelerating, because its velocity is changing (direction). The acceleration is toward the center of the circle and is called centripetal (center-seeking) acceleration.

An object must have an inward acceleration when traveling in a uniform circular motion at a constant speed. For a car, the friction of the tires supplies this acceleration. If a car hits an icy spot on a curved road, it slides outward, because the centripetal acceleration is not great enough to keep it in a circular motion.

In general, whenever an object moves in a circle of radius r with a constant speed v, the magnitude of the centripetal acceleration a is given by the formula:

$$a = \frac{v^2}{r} \quad \longrightarrow \quad ar = v^2 \quad \longrightarrow \quad \sqrt{ar} = v$$

1. Determine the speed of a car traveling on a circular road with a radius of 50 miles where the magnitude of the centripetal acceleration is 24.5 miles per hour squared.

2. Highway safety officials are trying to find a safe speed for a new section of highway that has a curve with a radius of 34 miles. They know a vehicle traveling where the magnitude of the centripetal acceleration is 73.53 miles per hour squared will slide off the road. Find this unsafe speed and estimate a speed that is safe for this section of highway.

3. Find the speed of a person standing on the equator. Earth has a diameter of about 8000 miles. The magnitude of the centripetal acceleration of Earth is approximately 0.11 feet per second squared. (*Hint:* Change the radius from miles to feet. 1 mile = 5280 feet.)

Challenge: Skills and Applications

For use with pages 511–516

Throughout this page, assume that all variables represent nonnegative real numbers.

In Exercises 1–8, simplify the expression.

1. $\sqrt{26} \cdot \sqrt{39}$

2. $\sqrt{34} \cdot \sqrt{323}$

3. $\dfrac{\sqrt{57}}{\sqrt{76}}$

4. $\dfrac{\sqrt{161}}{5\sqrt{368}}$

5. $5\sqrt{6} \cdot \sqrt{10} \cdot \sqrt{21}$

6. $7\sqrt{20} \cdot \sqrt{15} \cdot 2\sqrt{33}$

7. $\dfrac{\sqrt{pq}}{\sqrt{(p^2q)}}$

8. $\dfrac{\sqrt{(p^2q^4)}}{\sqrt{(pq)}}$

In Exercises 9–12, evaluate the radical expressions $\sqrt{(a + b)}$ and $\sqrt{a} + \sqrt{b}$ for the given values of a and b.

9. $a = 4, b = 9$

10. $a = 49, b = 0$

11. $a = 9, b = 16$

12. $a = 36, b = 9$

13. Based on your answers from Exercises 9–12, what can you conclude?

14. Try examples to make a similar conclusion about $\sqrt{(a - b)}$ and $\sqrt{a} - \sqrt{b}$.

In Exercises 15–17, use the fact that $a^{\frac{1}{2}}$ is defined as $\sqrt{a}$.

15. Simplify $\sqrt{a} \cdot \sqrt{a}$.

16. Simplify $a^{\frac{1}{2}} \cdot a^{\frac{1}{2}}$ using the product of powers property.

17. Based on the answers from Exercises 15 and 16, do you think the product of powers property holds for fractional exponents?

18. Simplify $\left(\sqrt{a}\right)^4$.

19. Simplify $\left(a^{\frac{1}{2}}\right)^4$ using the power of a power property.

20. Based on the answers from Exercises 18 and 19, do you think the power of a power property holds for fractional exponents?

TEACHER'S NAME _____ CLASS _____ ROOM _____ DATE _____

Lesson Plan

2-day lesson (See *Pacing the Chapter*, TE pages 500C–500D)   **For use with pages 517–525**

GOALS   1. **Sketch the graph of a quadratic function.**
2. **Use quadratic models in real-life settings.**

State/Local Objectives _____
_____

✓ **Check the items you wish to use for this lesson.**

STARTING OPTIONS
____ Homework Check: TE page 514; Answer Transparencies
____ Warm-Up or Daily Homework Quiz: TE pages 518 and 516, CRB page 36, or Transparencies

TEACHING OPTIONS
____ Motivating the Lesson: TE page 519
____ Concept Activity: SE page 517
____ Lesson Opener (Graphing Calculator): CRB page 37 or Transparencies
____ Graphing Calculator Activity with Keystrokes: CRB pages 38–42
____ Examples: Day 1: 1–2, SE page 519; Day 2: 3, SE page 520
____ Extra Examples: Day 1: TE page 519 or Transp.; Day 2: TE page 520 or Transp.
____ Technology Activity: SE page 525
____ Closure Question: TE page 520
____ Guided Practice: SE page 521; Day 1: Exs. 1–19; Day 2: Ex. 20

APPLY/HOMEWORK
Homework Assignment
____ Basic   Day 1: 24–64 multiples of 4; Day 2: 65, 66, 69–71, 76, 84, 86, 90, 94, 100; Quiz 1: 1–25
____ Average   Day 1: 24–64 multiples of 4; Day 2: 65, 66, 69–71, 76, 84, 86, 90, 94, 100; Quiz 1: 1–25
____ Advanced   Day 1: 24–64 multiples of 4; Day 2: 65, 66, 69–80, 84, 86, 90, 94, 100; Quiz 1: 1–25

Reteaching the Lesson
____ Practice Masters: CRB pages 43–45 (Level A, Level B, Level C)
____ Reteaching with Practice: CRB pages 46–47 or Practice Workbook with Examples
____ Personal Student Tutor

Extending the Lesson
____ Applications (Real-Life): CRB page 49
____ Challenge: SE page 523; CRB page 50 or Internet

ASSESSMENT OPTIONS
____ Checkpoint Exercises: Day 1: TE page 519 or Transp.; Day 2: TE page 520 or Transp.
____ Daily Homework Quiz (9.3): TE page 524, CRB page 54, or Transparencies
____ Standardized Test Practice: SE page 523; TE page 524; STP Workbook; Transparencies
____ Quiz (9.1–9.3): SE page 524; CRB page 51

Notes _____
_____
_____

TEACHER'S NAME _____ CLASS _____ ROOM _____ DATE _____

Lesson Plan for Block Scheduling

1-day lesson (See *Pacing the Chapter,* TE pages 500C–500D)   For use with pages 517–525

GOALS   1. **Sketch the graph of a quadratic function.**
            2. **Use quadratic models in real-life settings.**

State/Local Objectives _____

_____

_____

✓ **Check the items you wish to use for this lesson.**

STARTING OPTIONS
____ Homework Check: TE page 514; Answer Transparencies
____ Warm-Up or Daily Homework Quiz: TE pages 518 and
      516, CRB page 36, or Transparencies

TEACHING OPTIONS
____ Motivating the Lesson: TE page 519
____ Concept Activity: SE page 517
____ Lesson Opener (Graphing Calculator): CRB page 37 or Transparencies
____ Graphing Calculator Activity with Keystrokes: CRB pages 38–42
____ Examples: Day 2: 1–2, SE page 519; Day 3: 3, SE page 520
____ Extra Examples: Day 2: TE page 519 or Transp.; Day 3: TE page 520 or Transp.
____ Technology Activity: SE page 525
____ Closure Question: TE page 520
____ Guided Practice: SE page 521; Day 2: Exs. 1–19; Day 3: Ex. 20

APPLY/HOMEWORK
Homework Assignment (See also the assignments for Lessons 9.2 and 9.4.)
____ Block Schedule: Day 2: 24–64 multiples of 4
                   Day 3: 65, 66, 69–71, 76, 84, 86, 90, 94, 100; Quiz 1: 1–25

Reteaching the Lesson
____ Practice Masters: CRB pages 43–45 (Level A, Level B, Level C)
____ Reteaching with Practice: CRB pages 46–47 or Practice Workbook with Examples
____ Personal Student Tutor

Extending the Lesson
____ Applications (Real-Life): CRB page 49
____ Challenge: SE page 523; CRB page 50 or Internet

ASSESSMENT OPTIONS
____ Checkpoint Exercises: Day 2: TE page 519 or Transp.; Day 3: TE page 520 or Transp.
____ Daily Homework Quiz (9.3): TE page 524, CRB page 54, or Transparencies
____ Standardized Test Practice: SE page 523; TE page 524; STP Workbook; Transparencies
____ Quiz (9.1–9.3): SE page 524; CRB page 51

Notes _____

_____

_____

CHAPTER PACING GUIDE	
Day	**Lesson**
1	Assess Ch. 8; 9.1 (all)
2	9.2 (all); **9.3 (begin)**
3	**9.3 (end)**; 9.4 (all)
4	9.5 (all)
5	9.6 (all)
6	9.7 (all); 9.8 (begin)
7	9.8 (end); Review Ch. 9
8	Assess Ch. 9; 10.1 (all)

WARM-UP EXERCISES

For use before Lesson 9.3, pages 517–525

Evaluate $-\dfrac{b}{2a}$ for the following values.

1. $a = -3$ and $b = 8$

2. $a = 4$ and $b = -6$

Evaluate the expression for $x = -4$ and $x = 3$.

3. $x^2 + 4x - 1$

4. $-2x^2 - 7x + 3$

DAILY HOMEWORK QUIZ

For use after Lesson 9.2, pages 511–516

Simplify the expression.

1. $\sqrt{112}$

2. $-3\sqrt{\dfrac{32}{4}}$

3. $6\sqrt{\dfrac{28}{9}}$

4. $\dfrac{\sqrt{76}}{\sqrt{25}}$

5. $\sqrt{12} \cdot \dfrac{\sqrt{18}}{\sqrt{3}}$

6. A rectangle has width $2\sqrt{5}$ and length $\sqrt{20}$. Find its area.

Algebra 1
Chapter 9 Resource Book

Graphing Calculator Lesson Opener

For use with pages 518–524

1. Use your calculator to graph the function $y = x^2 + 1$.

   a. Describe the graph.

   b. Notice that your graph has one point that is lower than all the other points. Use the graph to estimate the coordinates of this point. Then use the TABLE feature of your calculator to check your estimate.

   c. Find and describe the vertical line that passes through the point you found in Step b. Describe the relationship of the graph of $y = x^2 + 1$ to this line.

2. Use your calculator to graph the function $y = 2x^2$.

   a. Describe the graph.

   b. Repeat Step 1b for this function.

   c. Find and describe the vertical line that passes through the point you found in Step b. Describe the relationship of the graph of $y = 2x^2$ to this line.

3. Use your calculator to graph the function $y = -x^2 + 2$.

   a. Describe the graph.

   b. Repeat Step 1b for this function except find the highest point.

   c. Find and describe the vertical line that passes through the point you found in Step b. Describe the relationship of the graph of $y = -x^2 + 2$ to this line.

NAME _____ DATE _____

Graphing Calculator Activity

For use with pages 518–525

GOAL To graphically determine the point(s) of intersection of two quadratic equations

In Chapter 7, you learned how to solve a system of linear equations. Finding the point(s) of intersection of two quadratic equations is solving a system of quadratic equations.

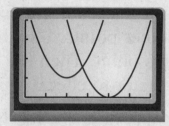

Activity

1 Enter and plot the pair of quadratic equations in the same coordinate plane.

$$y = x^2 - 8x + 19 \qquad\qquad y = -x^2 + 6x - 2$$

2 Use the Intersection feature of your graphing calculator to estimate the point(s) of intersection of the two quadratic equations in Step 1.

3 Repeat Steps 1 and 2 above for the pair of quadratic equations.

$$y = x^2 - 6x + 13$$
$$y = -x^2 - 4x + 2$$

Exercises

In Exercises 1–6, enter and plot the pair of quadratic equations in the same coordinate plane. Estimate the point(s) of intersection.

1. $y = x^2 - 7$
$y = -x^2 + 7$

2. $y = x^2 - 3x - 4$
$y = -x^2 - 4x + 5$

3. $y = x^2 + x + 2$
$y = x^2 - 2x + 6$

4. $y = x^2 + 4x + 9$
$y = -x^2 + 10x - 31$

5. $y = 2x^2 + 4x - 2$
$y = x^2 + 2x$

6. $y = 2x^2$
$y = -x^2 + 4x - 1$

7. How many point(s) of intersection are possible for two quadratic equations?

See page 39 for keystrokes.

NAME _____ DATE _____

Graphing Calculator Activity

For use with pages 518–525

TI-82

`Y=` `X,T,θ` `x²` `−` 8 `X,T,θ` `+` 19 `ENTER`

`(-)` `X,T,θ` `x²` `+` 6 `X,T,θ` `−` 2 `ENTER`

`ZOOM` 6 `2nd` [CALC] 5 `ENTER`

`ENTER`

Use the cursor keys, `◄` and `►`, to move the trace cursor to the point of intersection near $x = 2.1$. Press `ENTER`.

`2nd` [CALC] 5 `ENTER` `ENTER`

Use the cursor keys, `◄` and `►`, to move the trace cursor to the point of intersection near $x = 4.8$. Press `ENTER`.

`Y=` `CLEAR` `X,T,θ` `x²` `−` 6 `X,T,θ` `+` 13

`ENTER`

`CLEAR` `(-)` `X,T,θ` `x²` `−` 4 `X,T,θ` `+` 2

`ENTER`

`GRAPH`

TI-83

`Y=` `X,T,θ,n` `x²` `−` 8 `X,T,θ,n` `+` 19 `ENTER`

`(-)` `X,T,θ,n` `x²` `+` 6 `X,T,θ,n` `−` 2 `ENTER`

`ZOOM` 6 `2nd` [CALC] 5 `ENTER`

`ENTER` 2.1 `ENTER`

`2nd` [CALC] 5 `ENTER` `ENTER` 4.8 `ENTER`

`Y=` `CLEAR` `X,T,θ,n` `x²` `−` 6 `X,T,θ,n` `+` 13

`ENTER`

`CLEAR` `(-)` `X,T,θ,n` `x²` `−` 4 `X,T,θ,n` `+` 2

`ENTER`

`GRAPH`

SHARP EL-9600c

`Y=` `X/θ/T/n` `x²` `−` 8 `X/θ/T/n` `+` 19 `ENTER`

`(-)` `X/θ/T/n` `x²` `+` 6 `X/θ/T/n` `−` 2 `ENTER`

`ZOOM` [A] 5 `2ndF` [CALC] 2

`2ndF` [CALC] 2

`Y=` `CL` `X/θ/T/n` `x²` `−` 6 `X/θ/T/n` `+` 13

`ENTER`

`CL` `(-)` `X/θ/T/n` `x²` `−` 4 `X/θ/T/n` `+` 2

`ENTER`

`GRAPH`

CASIO CFX-9850GA PLUS

From the main menu, choose GRAPH.

`X,θ,T` `x²` `−` 8 `X,θ,T` `+` 19 `EXE`

`(-)` `X,θ,T` `x²` `+` 6 `X,θ,T` `−` 2 `EXE`

`SHIFT` `F3` `F3` `EXIT` `F6` `F5` `F5` `►`

`EXIT` `▲` `▲` `X,θ,T` `x²` `−` 6 `X,θ,T`

`+` 13

`EXE`

`(-)` `X,θ,T` `x²` `−` 4 `X,θ,T` `+` 2 `EXE`

`F6`

Graphing Calculator Activity Keystrokes

For use with Developing Concepts Activity 9.3 on page 517

TI-82

| Y= | (-) | 2 | X,T,θ | x^2 | ENTER |

| WINDOW | ZOOM | 6 |

TI-83

| Y= | (-) | 2 | X,T,θ,n | x^2 | ENTER |

| WINDOW | ZOOM | 6 |

SHARP EL-9600c

| Y= | (-) | 2 | X/θ/T/n | x^2 | ENTER |

| ZOOM | [A]5 |

CASIO CFX-9850GA PLUS

From the main menu, choose GRAPH.

| (-) | 2 | X,θ,T | x^2 | EXE |

| SHIFT | F3 | F3 | EXIT | F6 |

Algebra 1
Chapter 9 Resource Book

NAME _____ DATE _____

Graphing Calculator Activity Keystrokes

For use with Technology Activity 9.3 on page 525

TI-82

STAT 1

Enter *x*-values (time) in L1.

0 ENTER 0.04 ENTER 0.12 ENTER

0.16 ENTER 0.22 ENTER 0.26 ENTER

0.32 ENTER 0.36 ENTER 0.41 ENTER

Enter *y*-values (height) in L2.

5.23 ENTER 5.16 ENTER 4.85 ENTER

4.63 ENTER 4.2 ENTER 3.85 ENTER

3.23 ENTER 2.77 ENTER 1.98 ENTER

2nd [STAT PLOT] 1

Choose the following.

On; Type: ⌊∴ ; Xlist: L1; Ylist: L2; Mark: ▫

WINDOW ENTER 0 ENTER 1 ENTER 1

ENTER 0 ENTER 6 ENTER 1 ENTER

GRAPH

STAT ▶ 6 2nd [L1] , 2nd [L2] ENTER

Y= VARS 5 ▶ ▶ 7

GRAPH

TI-83

STAT 1

ENTER *x*-values (time) in L1.

0 ENTER 0.04 ENTER 0.12 ENTER

0.16 ENTER 0.22 ENTER 0.26 ENTER

0.32 ENTER 0.36 ENTER 0.41 ENTER

Enter *y*-values (height) in L2.

5.23 ENTER 5.16 ENTER 4.85 ENTER

4.63 ENTER 4.2 ENTER 3.85 ENTER

3.23 ENTER 2.77 ENTER 1.98 ENTER

2nd [STAT PLOT] 1

Choose the following.

On; Type: ⌊∴ ; Xlist: L1; Ylist: L2; Mark: ▫

WINDOW 0 ENTER 1 ENTER 1 ENTER

0 ENTER 6 ENTER 1 ENTER

GRAPH

STAT ▶ 5 2nd [L1] , 2nd [L2] ENTER

Y= VARS 5 ▶ ▶ 1

GRAPH

Keystrokes continued on next page.

Algebra 1
Chapter 9 Resource Book

Graphing Calculator Activity Keystrokes

For use with Technology Activity 9.3 on page 525

SHARP EL-9600c

`STAT` [A] `ENTER`

Enter *x*-values (time) in L1.

0 `ENTER` 0.04 `ENTER` 0.12 `ENTER`

0.16 `ENTER` 0.22 `ENTER` 0.26 `ENTER`

0.32 `ENTER` 0.36 `ENTER` 0.41 `ENTER`

Enter *y*-values (height) in L2.

5.23 `ENTER` 5.16 `ENTER` 4.85 `ENTER`

4.63 `ENTER` 4.2 `ENTER` 3.85 `ENTER`

3.23 `ENTER` 2.77 `ENTER` 1.98 `ENTER`

`2ndF` [STAT PLOT] [A] `ENTER`

Choose the following.

on; DATA XY; ListX: L1; ListY: L2

`2ndF` [STAT PLOT] [G] 3

`WINDOW` 0 `ENTER` 1 `ENTER` 1 `ENTER`

0 `ENTER` 6 `ENTER` 1 `ENTER`

`GRAPH`

`2ndF` [QUIT]

`STAT` [D] 0 4

`(` `2ndF` [L1] `,` `2ndF` [L2] `,` `VARS` [A]

`ENTER` [A] 1 `)` `ENTER`

`GRAPH`

CASIO CFX-9850Gᴀ PLUS

From the main menu, choose STAT.

Enter *x*-values (time) in List 1.

0 `EXE` 0.04 `EXE` 0.12 `EXE`

0.16 `EXE` 0.22 `EXE` 0.26 `EXE`

0.32 `EXE` 0.36 `EXE` 0.41 `EXE`

Enter *y*-values (height) in List 2.

5.23 `EXE` 5.16 `EXE` 4.85 `EXE`

4.63 `EXE` 4.2 `EXE` 3.85 `EXE`

3.23 `EXE` 2.77 `EXE` 1.98 `EXE`

`F1` `F6`

Choose the following.

Graph Type: Scatter; XList: List 1: YList: List 2;
Frequency: 1; Mark Type:▪

`EXIT`

`SHIFT` `F3` 0 `EXE` 20 `EXE` 1 `EXE` 0 `EXE` 1

`EXE` 0.1 `EXE` `EXIT`

`F1` `F1` `F3` `F6`

Algebra 1
Chapter 9 Resource Book

Practice A
For use with pages 518–524

Identify the values of *a*, *b*, and *c* in the functions.

1. $y = 3x^2 - 5x + 2$ **2.** $y = x^2 + 2x - 3$ **3.** $y = {}^-4x^2 + x$

4. $y = -x^2 + 4x - 8$ **5.** $y = -5x^2 - x + 5$ **6.** $y = x^2 - 4$

Tell whether the graph opens up or down. Write an equation of the axis of symmetry.

7. $y = 4x^2 - 4$ **8.** $y = x^2 - 2x - 3$ **9.** $y = -x^2 - 2x + 3$

10. $y = 5x^2 + 10x + 7$ **11.** $y = -x^2 + 4x + 16$ **12.** $y = -3x^2 - 9x - 12$

13. $y = -2x^2 - 3x + 6$ **14.** $y = 7x^2 + 14x - 2$ **15.** $y = 3x^2 + 2x + 4$

Find the coordinates of the vertex.

16. $y = 3x^2$ **17.** $y = -2x^2$ **18.** $y = 5x^2 - 1$

19. $y = x^2 + 6x$ **20.** $y = x^2 + 6x + 2$ **21.** $y = -2x^2 + 4x - 1$

Find the coordinates of the vertex. Make a table of values, using *x*-values to the left and right of the vertex.

22. $y = x^2 + 2x + 4$

x				
y				

23. $y = -3x^2 + 6x + 1$

x				
y				

Sketch the graph of the function. Label the vertex.

24. $y = -x^2 - 4$ **25.** $y = x^2 + 6x + 5$ **26.** $y = -x^2 - 4x - 3$

27. $y = x^2 + 2x - 15$ **28.** $y = 3x^2$ **29.** $y = x^2 - 6x + 10$

30. $y = -2x^2 - 8x + 20$ **31.** $y = -x^2 + 2x + 5$ **32.** $y = 2x^2 - 6x + 4$

33. *Gateway Arch* The Gateway Arch in St. Louis, Missouri, has a shape similar to that of a parabola. The edge of the arch can be modeled by

$$h = -\frac{2}{315}x^2 + 4x$$

where *x* and *h* are measured in feet. How high is the arch?

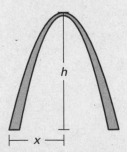

34. *Valley Depth* A model for a valley between two mountains whose peaks touch the *x*-axis is $y = 40.4x^2 - 404x$, where *x* and *y* are measured in feet. How deep is the valley?

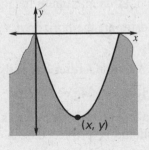

Practice B

For use with pages 518–524

Identify the values of *a*, *b*, and *c* in the function.

1. $y = -2x^2$

2. $y = x^2 - 9x + 5$

3. $y = 3x^2 + 7x$

4. $y = \frac{1}{2}x^2 - 2x - \frac{1}{4}$

5. $y = -4.5x^2 + 4$

6. $y = 1.7x^2 + 2.3x + 1.1$

Tell whether the graph opens up or down. Find the coordinates of the vertex. Write an equation of the axis of symmetry.

7. $y = 3x^2$

8. $y = -3x^2 + 8x$

9. $y = -4x^2 - 4x + 8$

10. $y = 2x^2 - 4x + 3$

11. $y = 3x^2 - 12x - 2$

12. $y = 2x^2 + 3x + 6$

13. $y = 2x^2 + 7x - 21$

14. $y = -\frac{1}{4}x^2 - 16$

15. $y = -6x^2 + 14x$

Find the coordinates of the vertex. Make a table of values using *x*-values to the left and right of the vertex.

16. $y = -x^2 - 2x + 15$

x					
y					

17. $y = 3x^2 + 2x + 4$

x					
y					

Sketch the graph of the function. Label the vertex.

18. $y = -x^2$

19. $y = -x^2 - 6x$

20. $y = -2x^2 + 2x - 4$

21. $y = x^2 - 4$

22. $y = -3x^2 + 6x + 2$

23. $y = x^2 + 4x + 7$

24. $y = 2x^2 - x - 1$

25. $y = -x^2 + x - \frac{1}{3}$

26. $y = \frac{1}{2}x^2 - 4x + 1$

Throwing a Ball In Exercises 27 and 28, use the following information.

The path of a ball thrown into the air from a height of 3 feet is given by $y = -\frac{1}{8}x^2 + x + 3$, where *y* is the height of the ball in feet at the horizontal distance of *x* feet from the thrower.

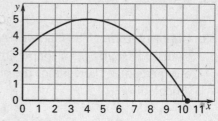

27. How high is the ball at its maximum height?

28. Make a table and estimate the horizontal distance the ball traveled before hitting the ground.

29. *Minimum Cost*   A manufacturer has daily production costs of

$$C = 10,000 - 10x + 0.045x^2$$

where *C* is the total cost in dollars and *x* is the number of units produced. How many units should be produced each day to yield a minimum cost?

NAME _____ DATE _____

Practice C

For use with pages 518–524

Tell whether the graph opens up or down. Find the coordinates of the vertex. Write an equation of the axis of symmetry.

1. $y = -9x^2$

2. $y = 8x^2 + 4x$

3. $y = -10x - 5x^2 + 3$

4. $y = -x^2 + 4x - 2$

5. $y = 2 - 10x^2$

6. $y = -6 + x^2 - 5x$

7. $y = 2x^2 - 5x - 6$

8. $y = 3x + 2 + 5x^2$

9. $y = 6 - x^2 + 7x$

Find the coordinates of the vertex. Make a table of values using *x*-values to the left and right of the vertex.

10. $y = 3x^2 + 6x - 2$

x					
y					

11. $y = 7x^2 + 2x - 10$

x					
y					

Sketch the graph of the function. Label the vertex.

12. $y = x^2 - 2x + 6$

13. $y = x^2 + 8x + 16$

14. $y = -x^2 + 6x - 9$

15. $y = -2x^2 - 2x - 3$

16. $y = 3x^2 + 2x + 4$

17. $y = x^2 - 4x + 7$

18. $y = -\frac{2}{5}x^2 + 4$

19. $y = \frac{1}{4}x^2 - 4$

20. $y = -\frac{1}{4}x^2 + 2x + 3$

21. $y = -\frac{1}{6}x^2 - 6$

22. $y = -\frac{1}{3}x^2 + 3x - 2$

23. $y = -\frac{3}{4}x^2 + 4x - 9$

Swish   **In Exercises 24 and 25, use the following information.**

In the diagram below, the backboard is located on the *y*-axis and the hoop is located at the point $(1, 10)$. A basketball thrown toward the hoop follows the path $y = -0.36x^2 + 2.8x + 7.56$, where *x* and *y* are measured in feet.

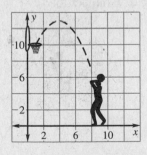

24. When the ball was at it highest point, what was its horizontal distance from the backboard?

25. At its highest point, how far off the ground was the basketball?

26. *Maximum Height of a Diver*   The path of a diver is given by

$$y = -\frac{4}{9}x^2 + \frac{24}{9}x + 10$$

where *y* is the height in feet and *x* is the horizontal distance from the end of the diving board in feet (see figure). Find the maximum height of the diver.

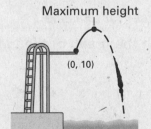

Maximum height

(0, 10)

NAME _____ DATE _____

Reteaching with Practice

For use with pages 518–524

GOAL Sketch the graph of a quadratic function and use quadratic models in real-life settings

> ### VOCABULARY
>
> A **quadratic function** is a function that can be written in the **standard form** $y = ax^2 + bx + c$, where $a \neq 0$.
>
> Every quadratic function has a U-shaped graph called a **parabola.**
>
> The **vertex** of a parabola is the lowest point of a parabola that opens up and the highest point of a parabola that opens down.
>
> The **axis of symmetry** of a parabola is the line passing through the vertex that divides the parabola into two symmetric parts.

EXAMPLE 1 *Sketching a Quadratic Function with a Positive a-value*

Sketch the graph of $y = x^2 - 2x + 1$.

SOLUTION

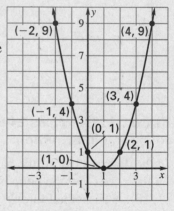

The vertex has an x-coordinate of $-\dfrac{b}{2a}$. Find the x-coordinate when $a = 1$ and $b = -2$.

$$-\frac{b}{2a} = -\frac{-2}{2(1)} = 1$$

Make a table of values, using x-values to the left and right of $x = 1$.

x	-2	-1	0	1	2	3	4
y	9	4	1	0	1	4	9

Plot the points. The vertex is $(1, 0)$ and the axis of symmetry is $x = 1$. Connect the points to form a parabola that opens up because a is positive.

Exercises for Example 1

Sketch the graph of the function. Label the vertex.

1. $y = 2x^2$ **2.** $y = x^2 + 3x$ **3.** $y = x^2 + 2x + 1$

NAME _____ DATE _____

Reteaching with Practice

For use with pages 518–524

EXAMPLE 2 *Sketching a Quadratic Function with a Negative a-value*

Sketch the graph of $y = -x^2 + 2x - 3$.

SOLUTION

The vertex has an x-coordinate of $-\dfrac{b}{2a}$. Find the
x-coordinate when $a = -1$ and $b = 2$.

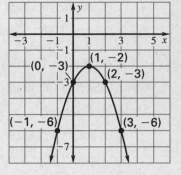

$$-\frac{b}{2a} = -\frac{2}{2(-1)} = 1$$

Make a table of values, using x-values to the left and right
of $x = 1$.

x	-1	0	1	2	3
y	-6	-3	-2	-3	-6

Plot the points. The vertex is $(1, -2)$ and the axis of symmetry is $x = 1$. Connect
the points to form a parabola that opens down because a is negative.

Exercises for Example 2

Sketch the graph of the function. Label the vertex.

4. $y = -4x^2$ **5.** $y = -x^2 + x$ **6.** $y = -x^2 - 2x + 3$

EXAMPLE 3 *Using a Quadratic Model*

A ball was thrown and followed a path described by $y = -0.02x^2 + x$.
What was the maximum height (in feet) of the thrown ball?

SOLUTION

The maximum height of the ball occurred at the vertex of the parabolic
path. Find the x-coordinate of the vertex. Use $a = -0.02$ and $b = 1$.

$$-\frac{b}{2a} = -\frac{1}{2(-0.02)} = \frac{1}{0.04} = 25$$

Substitute 25 for x in the model to find the maximum height.

$$y = -0.02(25)^2 + 25 = 12.5$$

The maximum height of the thrown ball was 12.5 feet.

Exercise for Example 3

7. Rework Example 3 if the path is described by $y = -0.01x^2 + x$.

Algebra 1
Chapter 9 Resource Book

Lesson 9.3

NAME _____ DATE _____

Quick Catch-Up for Absent Students

For use with pages 517–525

The items checked below were covered in class on (date missed) _____

Activity 9.3: Investigating Graphs of Quadratic Functions (p. 517)

_____ **Goal:** Determine how the coefficients a, b, and c affect the shape of the graph of the quadratic function $y = ax^2 + bx + c$.

_____ Student Help: Keystroke Help

Lesson 9.3: Graphing Quadratic Functions

_____ **Goal 1:** Sketch the graph of a quadratic function. (pp. 518–519)

Material Covered:

_____ Example 1: Graphing a Quadratic Function with a Positive a-Value

_____ Student Help: Study Tip

_____ Example 2: Graphing a Quadratic Function with a Negative a-Value

Vocabulary:

quadratic function, p. 518   standard form of a quadratic function, p. 518
parabola, p. 518   vertex, p. 518
axis of symmetry, p. 518

_____ **Goal 2:** Use quadratic models in real-life settings. (p. 520)

Material Covered:

_____ Example 3: Using a Quadratic Model

Activity 9.3: Graphing Quadratic Curves of Best Fit (p. 525)

_____ **Goal:** Find a quadratic model for data using a graphing calculator.

_____ Student Help: Keystroke Help

_____ Other (specify) _____

_____

Homework and Additional Learning Support

_____ Textbook (specify) pp. 521–524 _____

_____

_____ *Reteaching with Practice* worksheet (specify exercises)_____

_____ *Personal Student Tutor* for Lesson 9.3

LESSON 9.3

Real-Life Application: When Will I Ever Use This?

For use with pages 518–524

Ballet Recital

Your ballet class has decided to have a recital during December. The most popular choice of ballets voted on by class members is *The Nutcracker* by Pyotr (Peter) Tchaikovsky.

The music, completed in 1892, was commissioned by the Imperial Opera Directorate. Tchaikovsky actually disliked the subject matter of the ballet, which is a story based on a nutcracker and a mouse king. In spite of this, Tchaikovsky accepted the commission and the first performance of the ballet was on December 18, 1892. Critics did not like the finished work, but generations of ballet audiences have since disagreed with them. *The Nutcracker* is still one the most popular holiday ballets.

Your class chooses two primary dancers for the lead roles, one male and one female. One of the male dancer's leaps can be modeled by the equation $h = -8t^2 + 8t$ where h is height in feet and t is time in seconds. One of the female dancer's leaps can be modeled by the equation $h = -10t^2 + 8t$.

 1. Estimate the maximum height reached by the male dancer.

 2. How many seconds does it take the male dancer to reach his maximum jump height? Round to the nearest tenth.

 3. Estimate the maximum height reached by the female dancer. Round to the nearest tenth.

 4. How many seconds does it take the female dancer to reach her maximum height? Round to the nearest tenth.

 5. Sketch the graph of the equation for the male dancer.

Challenge: Skills and Applications

For use with pages 518–524

1. Sketch the graphs of $y = x^2$, $y = x^2 + 3$, and $y = x^2 - 2$ on the same coordinate grid.

2. Sketch the graphs of $y = -x^2$, $y = -x^2 + 4$, and $y = -x^2 - 1$ on the same coordinate grid.

3. Based on the graphs from Exercise 1, where is the graph of $y = x^2 + k$ in relation to the graph of $y = x^2$ for $k > 0$ and for $k < 0$?

4. Based on the graphs from Exercise 2, where is the graph of $y = -x^2 + k$ in relation to the graph of $y = -x^2$ for $k > 0$ and for $k < 0$?

5. Sketch the graphs of $y = x^2$, $y = (x - 4)^2$, and $y = (x + 2)^2$ on the same coordinate grid.

6. Based on the graphs from Exercise 5, where is the graph of $y = (x - h)^2$ in relation to the graph of $y = x^2$ for $h > 0$ and for $h < 0$?

In Exercises 7–9, find the value of a if the given point is on the graph of $y = ax^2$.

7. $(-3, 8)$ 8. $(2k, 7)$ 9. $(5k, 4k^2)$

In Exercises 10 and 11, find the values of a and b if both of the given points are on the graph of $y = ax^2 + b$.

Example: $(2, 11), (-3, 21)$

Solution: Substitute each ordered pair into $y = ax^2 + b$ to get a system of equations involving a and b:

$$11 = a(2)^2 + b \Rightarrow 11 = 4a + b$$

$$21 = a(-3)^2 + b \Rightarrow 21 = 9a + b$$

Solving the system, you get $(2, 3)$. Thus, $y = 2x^2 + 3$.

10. $(-2, 1), (4, 19)$ 11. $\left(3, -\frac{1}{2}\right), \left(-1, \frac{7}{2}\right)$

Algebra 1
Chapter 9 Resource Book

Lesson 9.3

NAME _____ DATE _____

Quiz 1

For use after Lessons 9.1–9.3

1. Evaluate the expression. *(Lesson 9.1)*

$$\frac{2 + 3\sqrt{49}}{5}$$

2. Solve the equation $3x^2 = 108$. *(Lesson 9.1)*

3. Simplify the expression $\sqrt{\dfrac{25}{49}}$. *(Lesson 9.2)*

4. Simplify the expression $\dfrac{\sqrt{28}}{5}$. *(Lesson 9.2)*

5. Tell whether the graph of $y = -x^2 - 4x - 1$ opens up or down. Find the coordinates of the vertex. Write the equation of the axis of symmetry of the function. *(Lesson 9.3)*

6. Sketch a graph of the function $y = -x^2 + 2x + 3$. *(Lesson 9.3)*

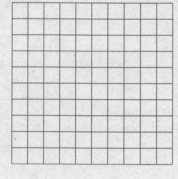

Answers

1. _____

2. _____

3. _____

4. _____

5. _____

6. Use grid at left. _____

TEACHER'S NAME _____ CLASS _____ ROOM _____ DATE _____

Lesson Plan

1-day lesson (See *Pacing the Chapter,* TE pages 500C–500D)   **For use with pages 526–532**

GOALS   1. **Solve a quadratic equation graphically.**
2. **Use quadratic models in real-life settings.**

State/Local Objectives _____

_____

✓ **Check the items you wish to use for this lesson.**

STARTING OPTIONS
____ Homework Check: TE page 521; Answer Transparencies
____ Warm-Up or Daily Homework Quiz: TE pages 526 and 524, CRB page 54, or Transparencies

TEACHING OPTIONS
____ Motivating the Lesson: TE page 527
____ Lesson Opener (Visual Approach): CRB page 55 or Transparencies
____ Graphing Calculator Activity with Keystrokes: CRB pages 56–58
____ Examples 1–4: SE pages 526–528
____ Extra Examples: TE pages 527–528 or Transparencies; Internet
____ Technology Activity: SE page 532
____ Closure Question: TE page 528
____ Guided Practice Exercises: SE page 529

APPLY/HOMEWORK
Homework Assignment
____ Basic   18–44 even, 52, 57, 58, 65, 72–82 even, 83
____ Average   18–44 even, 52, 57, 58, 65, 72–82 even, 83
____ Advanced   18–44 even, 52, 57–65, 72–82 even, 83

Reteaching the Lesson
____ Practice Masters: CRB pages 59–61 (Level A, Level B, Level C)
____ Reteaching with Practice: CRB pages 62–63 or Practice Workbook with Examples
____ Personal Student Tutor

Extending the Lesson
____ Applications (Interdisciplinary): CRB page 65
____ Challenge: SE page 531; CRB page 66 or Internet

ASSESSMENT OPTIONS
____ Checkpoint Exercises: TE pages 527–528 or Transparencies
____ Daily Homework Quiz (9.4): TE page 531, CRB page 69, or Transparencies
____ Standardized Test Practice: SE page 531; TE page 531; STP Workbook; Transparencies

Notes _____

_____

_____

TEACHER'S NAME _____ CLASS _____ ROOM _____ DATE _____

Lesson Plan for Block Scheduling

Half-day lesson (See *Pacing the Chapter,* TE pages 500C–500D)   For use with pages 526–532

GOALS
1. **Solve a quadratic equation graphically.**
2. **Use quadratic models in real-life settings.**

State/Local Objectives _____

_____

_____

✓ **Check the items you wish to use for this lesson.**

STARTING OPTIONS

____ Homework Check: TE page 521; Answer Transparencies

____ Warm-Up or Daily Homework Quiz: TE pages 526 and
     524, CRB page 54, or Transparencies

TEACHING OPTIONS

____ Motivating the Lesson: TE page 527

____ Lesson Opener (Visual Approach): CRB page 55 or Transparencies

____ Graphing Calculator Activity with Keystrokes: CRB pages 56–58

____ Examples 1–4: SE pages 526–528

____ Extra Examples: TE pages 527–528 or Transparencies; Internet

____ Technology Activity: SE page 532

____ Closure Question: TE page 528

____ Guided Practice Exercises: SE page 529

APPLY/HOMEWORK

Homework Assignment (See also the assignment for Lesson 9.3.)

____ Block Schedule: 18–44 even, 52, 57, 58, 65, 72–82 even, 83

Reteaching the Lesson

____ Practice Masters: CRB pages 59–61 (Level A, Level B, Level C)

____ Reteaching with Practice: CRB pages 62–63 or Practice Workbook with Examples

____ Personal Student Tutor

Extending the Lesson

____ Applications (Interdisciplinary): CRB page 65

____ Challenge: SE page 531; CRB page 66 or Internet

ASSESSMENT OPTIONS

____ Checkpoint Exercises: TE pages 527–528 or Transparencies

____ Daily Homework Quiz (9.4): TE page 531, CRB page 69, or Transparencies

____ Standardized Test Practice: SE page 531; TE page 531; STP Workbook; Transparencies

Notes _____

_____

_____

CHAPTER PACING GUIDE	
Day	**Lesson**
1	Assess Ch. 8; 9.1 (all)
2	9.2 (all); 9.3 (begin)
3	9.3 (end); **9.4 (all)**
4	9.5 (all)
5	9.6 (all)
6	9.7 (all); 9.8 (begin)
7	9.8 (end); Review Ch. 9
8	Assess Ch. 9; 10.1 (all)

NAME _____ DATE _____

WARM-UP EXERCISES

For use before Lesson 9.4, pages 526–532

Solve the equation.

1. $3x^2 = 75$

2. $x^2 + 11 = 36$

3. $-2x^2 + 8 = -154$

4. $\frac{1}{4}x^2 = 64$

....

DAILY HOMEWORK QUIZ

For use after Lesson 9.3, pages 517–525

Tell whether the graph opens up or down. Give the coordinates of the vertex and an equation of the axis of symmetry.

1. $y = 3x^2 - 3x + 4$

2. $y = -2x^2 - x + 4$

3. Graph $y = -2x^2 + 4x + 1$. Label the vertex.

4. A kangaroo's jump follows the path $h = -0.0347d^2 + 0.833d$, where h is the height in feet and d is the horizontal distance in feet. What is the maximum height that the kangaroo reaches?

NAME _____ DATE _____

Visual Approach Lesson Opener

For use with pages 526–531

Study the graph of the quadratic function. Then choose the correct solution to the equation. Explain how you used the graph to help you choose the correct solution.

1. $x^2 + x - 2 = 0$

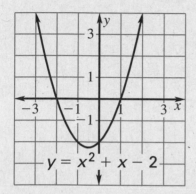

$y = x^2 + x - 2$

A. $x = -1$ or $x = 2$

B. $x = -2$ or $x = 1$

C. $x = 1$ or $x = 2$

D. $x = -2$ or $x = -1$

2. $x^2 - 2x - 8 = 0$

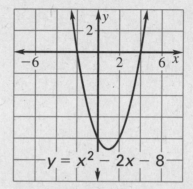

$y = x^2 - 2x - 8$

A. $x = -4$ or $x = 2$

B. $x = 2$ or $x = 4$

C. $x = -4$ or $x = -2$

D. $x = -2$ or $x = 4$

3. $x^2 - 4x + 3 = 0$

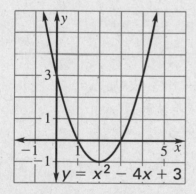

$y = x^2 - 4x + 3$

A. $x = 1$ or $x = 3$

B. $x = -1$ or $x = 3$

C. $x = -3$ or $x = -1$

D. $x = -3$ or $x = 1$

4. $x^2 + 3x + 2 = 0$

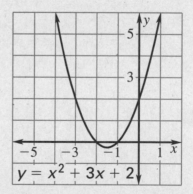

$y = x^2 + 3x + 2$

A. $x = -2$ or $x = 1$

B. $x = -1$ or $x = -2$

C. $x = -1$ or $x = 2$

D. $x = 1$ or $x = 2$

Graphing Calculator Activity Keystrokes

For use with page 530.

Keystrokes for Exercise 51

TI-82

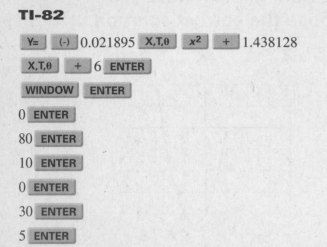

TI-83

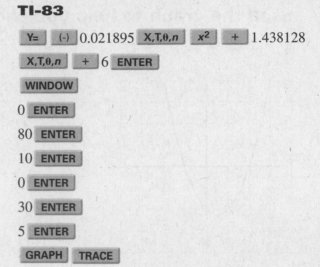

SHARP EL-9600c

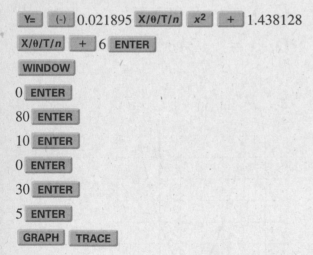

CASIO CFX-9850GA PLUS

From the main menu, choose RUN.

Y= (-) 0.021895 X,θ,T x² + 1.438128
X,θ,T + 6 EXE
SHIFT F3
0 EXE
80 EXE
10 EXE
0 EXE
30 EXE
5 EXE
EXIT
F6 F1

NAME _____ DATE _____

Graphing Calculator Activity Keystrokes

For use with Technology Activity 9.4 on page 532.

TI-82

Y= 2 X,T,θ x² + 3 X,T,θ − 4 ENTER

WINDOW

ZOOM 6

Find the postive root

Use the cursor keys, ◀ and ▶, to move the trace cursor to select the lower bound at $x \approx 0$. Press ENTER.

Use the cursor keys, ◀ and ▶, to move the trace cursor to select the upper bound at $x \approx 1.1$. Press ENTER.

Use the cursor keys, ◀ and ▶, to move the trace cursor to select the guess at $x \approx 0.85$. Press ENTER.

Find the negative root

Use the cursor keys, ◀ and ▶, to move the trace cursor to select the lower bound at $x \approx -2.6$. Press ENTER.

Use the cursor keys, ◀ and ▶, to move the trace cursor to select the upper bound at $x \approx 0$. Press ENTER.

Use the cursor keys, ◀ and ▶, to move the trace cursor to select the guess at $x \approx -2.3$. Press ENTER.

TI-83

Y= 2 X,T,θ,n x² + 3 X,T,θ,n − 4 ENTER

WINDOW

ZOOM 6

Find the postive root

2nd [CALC] 2 0 ENTER 1.1 ENTER 0.85 ENTER

Find the negative root

2nd [CALC] 2 (-) 2.6 ENTER 0 ENTER (-)

2.3 ENTER

Keystrokes continued on next page.

Algebra 1
Chapter 9 Resource Book

57

Lesson 9.4

Graphing Calculator Activity

For use with Technology Activity 9.4 on page 532.

SHARP EL-9600c

Y= 2 X/θ/T/n x² + 3 X/θ/T/n − 4

ENTER

ZOOM [A] 5

2ndF [CALC]5

2ndF [CALC]5

CASIO CFX-9850GA PLUS

From the main menu, choose GRAPH.

2 X,θ,T x² + 3 X,θ,T − 4 EXE

SHIFT F3 F3 EXIT

F6

F5 F1 ▶

Lesson 9.4

Write the equation in the form $ax^2 + bx + c = 0$.

1. $3x^2 = 7$ **2.** $-x^2 - 5x = 3$ **3.** $x^2 = 4x - 2$

4. $5 = 2x^2 - 4x$ **5.** $-x^2 = -3x$ **6.** $6x^2 - 4x = 8$

For each quadratic equation, use the graph to identify the roots of the equation.

7. $x^2 = 1$ **8.** $-x^2 + 4 = 0$ **9.** $-x^2 - x + 2 = 0$

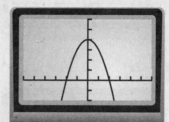 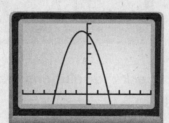 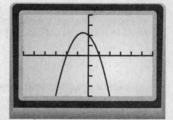

10. $x^2 - x - 12 = 0$ **11.** $-x^2 - x + 6 = 0$ **12.** $x^2 + 4x = 5$

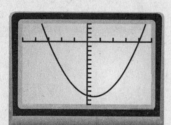 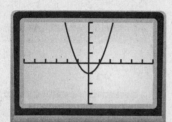 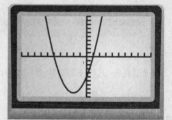

Solve the equation algebraically. Check the solutions graphically.

13. $2x^2 = 8$ **14.** $3x^2 = 27$ **15.** $\frac{1}{3}x^2 = 12$

16. $\frac{1}{2}x^2 = 32$ **17.** $x^2 + 2 = 27$ **18.** $x^2 - 9 = 40$

Solve the equation graphically. Check the solutions algebraically.

19. $x^2 - 4 = 0$ **20.** $x^2 - x = 2$ **21.** $x^2 - 3x = 4$

22. $-x^2 - x = -12$ **23.** $x^2 + 4 = 5x$ **24.** $-2x^2 - 4x = -6$

Ball Toss   **In Exercises 25–28, use the following information.**

The height h in feet of a ball t seconds after being tossed upward is given by the formula $h = 84t - 16t^2$.

25. Complete the following table of values.

t	0	1	2	3	4	5
h						

26. Sketch a graph of the model for positive values of x and y.

27. Use the graph to find a positive root of the equation $0 = 84t - 16t^2$.

28. After how many seconds will the ball hit the ground?

Practice B

For use with pages 526–531

For each quadratic equation, use the graph to identify the roots of the equation.

1. $x^2 = 4$

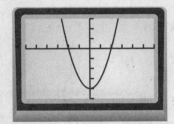

2. $-x^2 + 9 = 0$

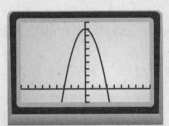

3. $-x^2 + 7x - 10 = 0$

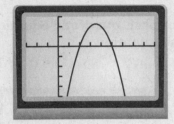

4. $2x^2 + 7x - 4 = 0$

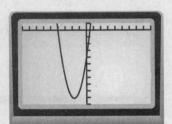

5. $-2x^2 + x + 10 = 0$

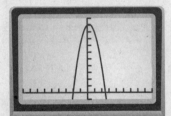

6. $\frac{1}{2}x^2 - x = \frac{15}{2}$

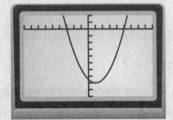

Solve the equation algebraically. Check the solutions graphically.

7. $2x^2 = 18$

8. $-2x^2 = -50$

9. $4x^2 = 36$

10. $\frac{1}{2}x^2 = 8$

11. $\frac{1}{3}x^2 = 27$

12. $\frac{2}{3}x^2 = 24$

13. $x^2 - 24 = 57$

14. $x^2 + 35 = 179$

15. $2x^2 - 5 = 27$

Solve the equations graphically. Check the solutions algebraically.

16. $-x^2 + x = -6$

17. $x^2 + 2x = 3$

18. $-x^2 + 6x = 8$

19. $x^2 + 5x = -6$

20. $4x^2 - 2 = 2x$

21. $2x^2 - 20 = -6x$

Ball Toss   **In Exercises 22–25, use the following information.**

The path of a ball thrown into the air from a height of 6 feet is given by

$$y = -\frac{1}{4}x^2 + \frac{5}{2}x + 6$$

where y is the height of the ball in feet at a horizontal distance of x feet from the thrower.

22. Complete the following table of values.

x	0	1	2	3	4	5	6	7
y								

23. Sketch a graph of the model for positive values of x and y.

24. Use the graph to find a positive root of the equation $0 = -\frac{1}{4}x^2 + \frac{5}{2}x + 6$.

25. Approximately how far from the thrower will the ball hit the ground?

NAME _____ DATE _____

Practice C

For use with pages 526–531

For each quadratic equation, use the graph to identify the roots of the equation.

1. $2x^2 + 2x - 4 = 0$ **2.** $x^2 - 6x + 9 = 0$ **3.** $x^2 - 5x = 0$

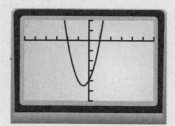   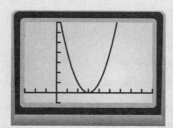   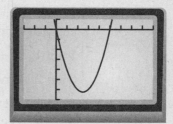

Solve the equation algebraically. Check the solutions graphically.

4. $x^2 = 100$ **5.** $3x^2 = 108$ **6.** $5x^2 = 245$

7. $\frac{1}{2}x^2 = 72$ **8.** $\frac{1}{3}x^2 = 75$ **9.** $\frac{3}{4}x^2 = 12$

10. $x^2 + 29 = 198$ **11.** $2x^2 - 14 = 114$ **12.** $3x^2 + 32 = 464$

Solve the equations graphically. Check the solutions algebraically.

13. $2x^2 - 4x + 2 = 0$ **14.** $x^2 + 12x + 36 = 0$ **15.** $x^2 + 3x = 28$

16. $3x^2 - 9x = 12$ **17.** $-x^2 + 24 = -10x$ **18.** $2x^2 + 11x = -12$

19. $-x^2 + 10x = 16$ **20.** $2x^2 - 12 = 5x$ **21.** $6x^2 + 9x = 6$

Use a graphing calculator to approximate the solutions of the equation.

22. $x^2 - 2x - 15 = 0$ **23.** $x^2 + 5x - 14 = 0$ **24.** $2x^2 - 2x - 24 = 0$

25. $-4x^2 - 4x + 8 = 0$ **26.** $\frac{1}{2}x^2 + x - 24 = 0$ **27.** $\frac{1}{4}x^2 + \frac{1}{4}x - 5 = 0$

Gateway Arch   **In Exercises 28–30, use the following information.**

The Gateway Arch in St. Louis, Missouri, was designed in the shape of a catenary (a U-shaped curve that resembles a parabola). The outer edge of the arch can be approximated by the equation

$$y = -0.006x^2 + 4.381x - 125.714$$

where x and y are measured in feet.

28. Sketch a graph of the model for positive values of x and y.

29. Use the graph to find the positive roots of the equation
   $0 = -0.006x^2 + 4.381x - 125.714.$

30. According to the model, how wide is the arch?

NAME _____ DATE _____

Reteaching with Practice

For use with pages 526–531

GOAL Solve a quadratic equation graphically and use quadratic models in real-life settings

---

VOCABULARY

The solution of a quadratic equation in one variable x can be solved or checked graphically with the following steps.

Step 1: Write the equation in the form $ax^2 + bx + c = 0$.

Step 2: Write the related function $y = ax^2 + bx + c$.

Step 3: Sketch the graph of the function $y = ax^2 + bx + c$. The solutions, or **roots**, of $ax^2 + bx + c = 0$ are the x-intercepts.

EXAMPLE 1 *Checking a Solution Using a Graph*

a. Solve $3x^2 = 75$ algebraically.   **b.** Check your solution graphically.

SOLUTION

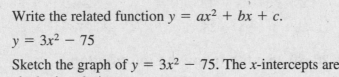

a. $3x^2 = 75$ Write original equation.

   $x^2 = 25$ Divide each side by 3.

   $x = \pm 5$ Find the square root of each side.

b. Write the equation in the form $ax^2 + bx + c = 0$.

   $3x^2 = 75$ Write original equation.

   $3x^2 - 75 = 0$ Subtract 75 from each side.

   Write the related function $y = ax^2 + bx + c$.

   $y = 3x^2 - 75$

Sketch the graph of $y = 3x^2 - 75$. The x-intercepts are ± 5, which agrees with the algebraic solution.

Exercises for Example 1

Solve the equation algebraically. Check the solutions graphically.

1. $\frac{1}{3}x^2 = 12$ **2.** $3x^2 + 2 = 50$ **3.** $x^2 - 7 = 2$

EXAMPLE 2 *Solving an Equation Graphically*

a. Solve $x^2 - 3x = 4$ graphically.

b. Check your solution algebraically.

Reteaching with Practice

For use with pages 526–531

SOLUTION

a. Write the equation in the form $ax^2 + bx + c = 0$.

$\quad x^2 - 3x = 4$ Write original equation.

$x^2 - 3x - 4 = 0$ Subtract 4 from each side.

Write the related function $y = ax^2 + bx + c$.

$y = x^2 - 3x - 4$

Sketch the graph of the function $y = x^2 - 3x - 4$.

From the graph, the x-intercepts appear to be $x = -1$ and $x = 4$.

b. You can check your solution algebraically by substitution.

Check $x = -1$:   Check $x = 4$:

$\quad\quad x^2 - 3x = 4 \quad\quad\quad\quad x^2 - 3x = 4$

$(-1)^2 - 3(-1) \overset{?}{=} 4 \quad\quad 4^2 - 3(4) \overset{?}{=} 4$

$\quad\quad 1 + 3 = 4 \quad\quad\quad\quad 16 - 12 = 4$

Exercises for Example 2

Solve the equation graphically. Check the solutions algebraically.

4. $x^2 + x = 12$ **5.** $x^2 - 5x = -6$ **6.** $x^2 - 5x = 6$

EXAMPLE 3   Using Quadratic Equations in Real Life

The average cost of a license and registration for an automobile in the United States from 1991 through 1997 can be modeled by

$$y = -0.63x^2 + 15.08x + 151.57$$

where y represents the average cost of a license and registration. Let x be the number of years since 1990. Use the graph of the model to estimate the average cost of a license and registration in 1995.

License and Registration Costs

SOLUTION

The year 1995 corresponds to $x = 5$. From the graph of the quadratic equation, the average cost of a license and registration appears to be about 210 dollars.

Exercise for Example 3

7. Algebraically check the solution in Example 3.

NAME _____ DATE _____

Quick Catch-Up for Absent Students

For use with pages 526–532

The items checked below were covered in class on (date missed) _____

Lesson 9.4: Solving Quadratic Equations by Graphing

____ **Goal 1:** Solve a quadratic equation graphically. (pp. 526–527)

Material Covered:

    ____ Example 1: Checking a Solution Using a Graph

    ____ Example 2: Solving an Equation Graphically

    ____ Student Help: Look Back

    ____ Example 3: Using a Graphing Calculator

Vocabulary:

    roots, p. 526

____ **Goal 2:** Use quadratic models in real-life settings. (p. 528)

Material Covered:

    ____ Example 4: Comparing Two Quadratic Models

Activity 9.4: Solving Quadratic Equations by Graphing (p. 532)

____ **Goal:** Use the root or zero feature of a graphing calculator to approximate solutions, or roots, of an equation.

    ____ Student Help: Keystroke Help

____ Other (specify) _____

_____

Homework and Additional Learning Support

    ____ Textbook (specify) pp. 529–531 _____

    _____

    ____ Internet: Extra Examples at www.mcdougallittel.com

    ____ *Reteaching with Practice* worksheet (specify exercises)_____

    ____ *Personal Student Tutor* for Lesson 9.4

Interdisciplinary Applications

For use with pages 526–531

Air Pollution

EARTH SCIENCE Air pollution occurs when wastes dirty the air. People produce most of the wastes that cause air pollution. Such wastes can be in the form of gases or particles of solid or liquid matter.

Industrial processes produce a wide range of pollutants. Plants that produce plastic foams are a major source of chlorofluorocarbons (CFC's), compounds of chlorine, fluorine, and carbon. Many countries, including the United States, have agreed to end production of CFC's.

A chlorofluorocarbon is any one of a group of synthetic organic compounds that contain chlorine, fluorine, and carbon. They are used as refrigerants in air conditioners and refrigerators. CFC's are also used to make plastic foams for furniture and insulation. Scientific studies indicate that CFC's harm the environment by breaking down ozone molecules in the earth's upper atmosphere.

In Exercises 1 and 2, use the following information.

Your earth science class is doing a study on the emission of greenhouse gases. The amount of chlorofluorocarbons in the atmosphere from 1990 to 1996 can be modeled by $C = .3095t^2 - 26.9285t + 224.9$, where C is the number of chlorofluorocarbons in thousands of metric tons and t is the number of years with $t = 0$ representing 1990.

1. Sketch a graph of the model.

2. According to the model, in what year will the number of chlorofluorocarbons in the atmosphere be zero?

In Exercises 3 and 4, use the following information.

Hydrochlorofluorocarbons (HCFC's) are compounds containing hydrogen, chlorine, fluorine, and carbon. Industry and the scientific community view certain chemicals within this class of compounds to be acceptable alternatives to chlorofluorocarbons. The HCFC's have shorter atmospheric lifetimes than the CFC's and a much smaller capacity to harm the stratosphere where the ozone layer is found. Because they still contain chlorine and have the potential to destroy stratospheric ozone, they are viewed only as temporary replacements for the CFC's. The amount of hydrochlorofluorocarbons in the atmosphere from 1990 to 1996 can be modeled by $H = -.9524t^2 + 17.5t + 73$, where H is the number of hydrochlorofluorocarbons in thousands of metric tons and t is the number of years with $t = 0$ representing 1990.

3. Sketch a graph of the model over your graph from Exercise 1.

4. Because HCFC's are new there was a point in time when there were no HCFC's in the atmosphere. Find this year.

Challenge: Skills and Applications

For use with pages 526–531

In Exercises 1–4, write an equation with the given x-intercepts p and q whose graph passes through the given point.

Example: $p = 1, q = -2, (3, 20)$

Solution: An equation with x-intercepts at p and q has the form
$y = a(x - p)(x - q)$. Thus, $y = a(x - 1)(x + 2)$. To find the value
of a, substitute the given coordinates for x and y:

$20 = a(3 - 1)(3 + 2)$

$20 = 10a$

$2 = a$

The equation is $y = 2(x - 1)(x + 2)$.

1. $p = 3, q = -1, (4, -15)$ 2. $p = 3, q = 5, (1, -2)$

3. $p = 4, q = -1, (2, 30)$ 4. $p = -5, q = -2, \left(\frac{2}{3}, \frac{8}{9}\right)$

5. Find a quadratic equation whose graph has x-intercepts 5 and -3 and whose vertex has y-coordinate -8. (*Hint:* Sketch the graph.)

In Exercises 6 and 7, use the following information.

The arch of a stone bridge has the shape of a parabola that is 30 feet wide at the water line and rises to its highest point 12 feet above the water. Suppose the vertex of the parabola is at the origin of a coordinate system.

6. Write an equation for the parabola.

7. If the water level rises to 4 feet below the highest point of the bridge, how wide is the parabola at the water line?

In Exercises 8 and 9, use the following information.

In a physics experiment, a ball bearing rolling horizontally off the edge of a desktop 27 inches high follows a parabolic path, landing on the floor 18 inches from the base of the desk.

8. Write an equation for the path of the ball bearing. Use the point where the ball bearing rolls off the desktop as the origin of a coordinate system.

9. How much higher does the desktop need to be in order for the ball bearing to land 30 inches from the base of the desk?

Lesson 9.4

TEACHER'S NAME _____ CLASS _____ ROOM _____ DATE _____

Lesson Plan

2-day lesson (See *Pacing the Chapter*, TE pages 500C–500D)   **For use with pages 533–539**

GOALS   1. **Use the quadratic formula to solve a quadratic equation.**
          2. **Use quadratic models in real-life situations.**

State/Local Objectives _____

_____

✓ Check the items you wish to use for this lesson.

STARTING OPTIONS
____ Homework Check: TE page 529; Answer Transparencies
____ Warm-Up or Daily Homework Quiz: TE pages 533 and 531, CRB page 69, or Transparencies

TEACHING OPTIONS
____ Motivating the Lesson: TE page 534
____ Lesson Opener (Activity): CRB page 70 or Transparencies
____ Graphing Calculator Activity with Keystrokes: CRB pages 71–72
____ Examples: Day 1: 1–2, SE pages 533–534; Day 2: 3–4, SE pages 534–535
____ Extra Examples: Day 1: TE page 534 or Transp.; Day 2: TE pages 534–535 or Transp.; Internet
____ Technology Activity: SE page 539
____ Closure Question: TE page 535
____ Guided Practice: SE page 536; Day 1: Exs. 1–15; Day 2: Exs. 16–22

APPLY/HOMEWORK
Homework Assignment
____ Basic   Day 1: 24–52 even; Day 2: 54–80 even, 84, 87–90, 92–102 even
____ Average   Day 1: 24–52 even; Day 2: 54–80 even, 82–84, 87–90, 92–102 even
____ Advanced   Day 1: 24–52 even; Day 2: 54–80 even, 82–90, 92–102 even

Reteaching the Lesson
____ Practice Masters: CRB pages 73–75 (Level A, Level B, Level C)
____ Reteaching with Practice: CRB pages 76–77 or Practice Workbook with Examples
____ Personal Student Tutor

Extending the Lesson
____ Applications (Interdisciplinary): CRB page 79
____ Challenge: SE page 538; CRB page 80 or Internet

ASSESSMENT OPTIONS
____ Checkpoint Exercises: Day 1: TE page 534 or Transp.; Day 2: TE pages 534–535 or Transp.
____ Daily Homework Quiz (9.5): TE page 538, CRB page 83, or Transparencies
____ Standardized Test Practice: SE page 538; TE page 538; STP Workbook; Transparencies

Notes _____

_____

_____

TEACHER'S NAME _____ CLASS _____ ROOM _____ DATE _____

Lesson Plan for Block Scheduling

1-day lesson (See *Pacing the Chapter*, TE pages 500C–500D)   **For use with pages 533–539**

GOALS   1. **Use the quadratic formula to solve a quadratic equation.**
2. **Use quadratic models in real-life situations.**

State/Local Objectives _____

_____

✓ **Check the items you wish to use for this lesson.**

STARTING OPTIONS
____ Homework Check: TE page 529; Answer Transparencies
____ Warm-Up or Daily Homework Quiz: TE pages 533 and
        531, CRB page 69, or Transparencies

TEACHING OPTIONS
____ Motivating the Lesson: TE page 534
____ Lesson Opener (Activity): CRB page 70 or Transparencies
____ Graphing Calculator Activity with Keystrokes: CRB pages 71–72
____ Examples 1–4: SE pages 533–535
____ Extra Examples: TE pages 534–535 or Transparencies; Internet
____ Technology Activity: SE page 539
____ Closure Question: TE page 535
____ Guided Practice Exercises: SE page 536

APPLY/HOMEWORK
Homework Assignment
____ Block Schedule: 24–80 even, 82–84, 87–90, 92–102 even

Reteaching the Lesson
____ Practice Masters: CRB pages 73–75 (Level A, Level B, Level C)
____ Reteaching with Practice: CRB pages 76–77 or Practice Workbook with Examples
____ Personal Student Tutor

Extending the Lesson
____ Applications (Interdisciplinary): CRB page 79
____ Challenge: SE page 538; CRB page 80 or Internet

ASSESSMENT OPTIONS
____ Checkpoint Exercises: TE pages 534–535 or Transparencies
____ Daily Homework Quiz (9.5): TE page 538, CRB page 83, or Transparencies
____ Standardized Test Practice: SE page 538; TE page 538; STP Workbook; Transparencies

Notes _____

_____

_____

| CHAPTER PACING GUIDE ||
Day	Lesson
1	Assess Ch. 8; 9.1 (all)
2	9.2 (all); 9.3 (begin)
3	9.3 (end); 9.4 (all)
4	**9.5 (all)**
5	9.6 (all)
6	9.7 (all); 9.8 (begin)
7	9.8 (end); Review Ch. 9
8	Assess Ch. 9; 10.1 (all)

LESSON

9.5

NAME _____ DATE _____

WARM-UP EXERCISES

For use before Lesson 9.5, pages 533–539

Evaluate the expression $\sqrt{b^2 - 4ac}$ for the given values. Give the answer in simplest form.

1. $a = 3, b = 8, c = 0$ **2.** $a = -2, b = 0, c = 5$

3. $a = 2, b = 6, c = 3$ **4.** $a = -3, b = 8, c = 1$

5. $a = -3, b = 5, c = 0$

···

DAILY HOMEWORK QUIZ

For use after Lesson 9.4, pages 526–532

1. Use the graph of $y = x^2 + 0.5x - 1.5$
to identify the roots of $x^2 + 0.5x - 1.5 = 0$.

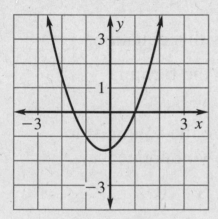

Solve the equation algebraically. Check the solutions graphically.

2. $x^2 + 5 = 21$ **3.** $3x^2 - 7 = 68$

Solve the equation graphically. Check the solutions algebraically.

4. $x^2 + 3x = 4$ **5.** $-3x^2 - 9x = -30$

Activity Lesson Opener

For use with pages 533–538

SET UP: Work with a partner.

For Questions 1–3, use the formula below.

$$\frac{-b \pm \sqrt{b^2 - 4ac}}{2a}$$

1. Consider the quadratic equation $x^2 - 3x - 10 = 0$.

   a. The equation is given in standard form, $ax^2 + bx + c = 0$. Name a, b, and c for this equation.

   b. Substitute the values for a, b, and c into the formula above. Simplify.

   c. Solve the equation by graphing. Compare the solutions to the values you found in Step b. What do you notice?

2. Consider the quadratic equation $x^2 + 3x - 4 = 0$.

   a. The equation is given in standard form, $ax^2 + bx + c = 0$. Name a, b, and c for this equation.

   b. Substitute the values for a, b, and c into the formula above. Simplify.

   c. Solve the equation by graphing. Compare the solutions to the values you found in Step b. What do you notice?

3. Consider the quadratic equation $x^2 + 6x + 8 = 0$.

   a. The equation is given in standard form, $ax^2 + bx + c = 0$. Name a, b, and c for this equation.

   b. Substitute the values for a, b, and c into the formula above. Simplify.

   c. Solve the equation by graphing. Compare the solutions to the values you found in Step b. What do you notice?

4. Make a conjecture about the expression given in Step b in Questions 1–3.

NAME _____ DATE _____

Graphing Calculator Activity Keystrokes

For use with page 539.

Keystrokes for Exercise 84

TI-82

| Y= | | (-) | 16 | X,T,θ | | x² | | + | 80 | X,T,θ | | – |

4 ENTER

WINDOW ENTER

0 ENTER

10 ENTER

1 ENTER

0 ENTER

110 ENTER

10 ENTER

GRAPH

TRACE

Move the trace cursor to the highest point on the graph.

ZOOM 2 ENTER

TRACE

TI-83

| Y= | | (-) | 16 | X,T,θ,n | | x² | | + | 80 | X,T,θ,n | | – |

4 ENTER

WINDOW

0 ENTER

10 ENTER

1 ENTER

0 ENTER

110 ENTER

10 ENTER

GRAPH

TRACE

Move the trace cursor to the highest point on the graph.

ZOOM 2 ENTER

TRACE

SHARP EL-9600c

| Y= | | (-) | 16 | X/θ/T/n | | x² | | + | 80 | X/θ/T/n | | – |

4 ENTER

WINDOW

0 ENTER

10 ENTER

1 ENTER

0 ENTER

110 ENTER

10 ENTER

GRAPH

TRACE

Move the trace cursor to the highest point on the graph.

ZOOM 2 ENTER

TRACE

CASIO CFX-9850GA PLUS

From the main menu, choose GRAPH.

| (-) | 16 | X,θ,T | | x² | | + | 80 | X,θ,T | | – |

4 EXE

SHIFT F3

0 EXE

10 EXE

1 EXE

0 EXE

110 EXE

10 EXE

EXIT F6

F1

Move the trace cursor to the highest point on the graph.

F2 F3 SHIFT F1

LESSON 9.5 CONTINUED

Graphing Calculator Activity Keystrokes

For use with Technology Activity 9.5 on page 539.

TI-82

PROGRAM: QUADFORM

:Prompt A, B, C

:$B^2 - 4AC \rightarrow D$

:If $D < 0$

:Then

:Disp "NO SOLUTION"

:Pause

:Else

:Disp "THE FIRST"

:Disp "SOLUTION IS. . ."

:Disp $(-B + \sqrt{D})/(2A)$

:Pause

:Disp "THE SECOND"

:Disp "SOLUTION IS. . ."

:Disp $(-B - \sqrt{D})/(2A)$

:Pause

:End

TI-83

PROGRAM: QUADFORM

:Prompt A, B, C

:$B^2 - 4AC \rightarrow D$

:If $D < 0$

:Then

:Disp "NO SOLUTION"

:Pause

:Else

:Disp "THE FIRST"

:Disp "SOLUTION IS. . ."

:Disp $(-B + \sqrt{D})/(2A)$

:Pause

:Disp "THE SECOND"

:Disp "SOLUTION IS. . ."

:Disp $(-B - \sqrt{D})/(2A)$

:Pause

:End

SHARP EL-9600c

QUADFORM

Input A

Input B

Input C

$B^2 - 4AC \Rightarrow D$

If $D < 0$ Goto 1

If $D \geq 0$ Goto 2

Label 1

Print "NO SOLUTION"

Wait

End

Label 2

Print "THE FIRST SOLUTION IS. . ."

Print $(-B + \sqrt{D})/(2A)$

Wait

Print "THE SECOND SOLUTION IS. . ."

Print $(-B - \sqrt{D})/(2A)$

Wait

End

CASIO CFX-9850GA PLUS

QUADFORM

"A="? $\rightarrow$ A↵

"B="? $\rightarrow$ B↵

"C="? $\rightarrow$ C↵

$B^2 - 4AC \rightarrow D$↵

If $D < 0$↵

Then "NO SOLUTION"↵

Else "THE FIRST SOLUTION IS. . ."↵

$(-B + \sqrt{D})/(2A)$↵

"THE SECOND SOLUTION IS. . ."↵

$(-B - \sqrt{D})/(2A)$↵

IfEnd↵

Practice A

For use with pages 533–538

Write the equation in standard form.

1. $4x^2 = 12$

2. $-3x^2 - 8x = 2$

3. $x^2 = 10x - 6$

4. $5x - 4 = 3x^2$

5. $4x^2 = -5x$

6. $8x^2 + 5x = 1$

Find the value of $b^2 - 4ac$ for the equation.

7. $x^2 - 2x - 2 = 0$

8. $x^2 - 6x - 4 = 0$

9. $3x^2 - 2x - 8 = 0$

10. $2x^2 + 6x + 3 = 0$

11. $2x^2 + 3x - 1 = 0$

12. $4x^2 + 4x - 1 = 0$

Use the quadratic formula to solve the equation.

13. $x^2 - 8x + 15 = 0$

14. $x^2 + 11x + 18 = 0$

15. $x^2 + 5x - 14 = 0$

16. $x^2 + x - 2 = 0$

17. $2x^2 + 3x - 2 = 0$

18. $4x^2 - 7x + 3 = 0$

19. $4x^2 + 3x - 1 = 0$

20. $x^2 - 2x - 8 = 0$

21. $x^2 + 2x - 2 = 0$

Find the x-intercepts of the graph of the equation.

22. $y = x^2 + 5x + 6$

23. $y = 2x^2 - 5x - 12$

24. $y = 5x^2 - 4x - 33$

25. $y = 6x^2 + 7x - 5$

26. $y = -x^2 + 9x - 18$

27. $y = -3x^2 + 7x - 4$

28. **Surface Area** The surface area of a rectangular box with a square base is 112 square inches. The surface area is given by $A = 2x^2 + 4xh$. Find x.

29. **Model Rocket** You launch a model rocket from ground level with an initial velocity of 95 feet per second. After how many seconds will the rocket have an altitude of 114 feet?

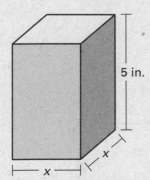

5 in.

x

x

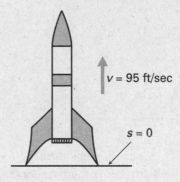

$v = 95$ ft/sec

$s = 0$

30. **Baseball** You throw a baseball in the air with a starting velocity of 28 feet per second. The baseball is 5 feet high when it leaves your hand. After how many seconds will it hit the ground? Use the vertical motion model to answer the question.

NAME _____ DATE _____

Practice B

For use with pages 533–538

Find the value of $b^2 - 4ac$ for the equation.

1. $3x^2 - 8x - 1 = 0$
2. $2x^2 + 5x - 2 = 0$
3. $15x^2 - 10x + 1 = 0$
4. $4x^2 + x - 2 = 0$
5. $x^2 - 6x + 4 = 0$
6. $5x^2 - 12x + \frac{1}{2} = 0$

Use the quadratic formula to solve the equation.

7. $4x^2 - 13x + 3 = 0$
8. $2x^2 + 7x + 3 = 0$
9. $-x^2 + x + 30 = 0$
10. $3x^2 + 7x - 20 = 0$
11. $-4x^2 + x + 14 = 0$
12. $2x^2 - x - 2 = 0$
13. $-2x^2 + 3x - 1 = 0$
14. $2x^2 + 10x - 5 = 0$
15. $-6x^2 + 4x - \frac{1}{3} = 0$
16. $-5x^2 + 7x - 2 = 0$
17. $8x^2 - 5x - 2 = 0$
18. $7x^2 + 9x - 1 = 0$

Find the x-intercepts of the graph of the equation.

19. $y = x^2 + 2x - 8$
20. $y = 2x^2 - 5x - 3$
21. $y = 6x^2 - x - 12$
22. $y = 3x^2 + 5x + 1$
23. $y = -5x^2 + 4x + 2$
24. $y = -2x^2 - 7x - 1$

25. **Chemistry Experiment** During a chemistry experiment, the cork in a 0.5-foot tall beaker with an effervescent solution pops off with an initial velocity of 20 feet per second. How many seconds does it take for the cork to hit the table?

26. **Fireworks** Fireworks are shot upward with an initial velocity of 125 feet per second from a platform 3 feet above the ground. Use the model $h = -16t^2 + vt + s$ to find how long it will take the rocket to hit the ground.

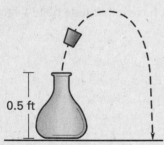

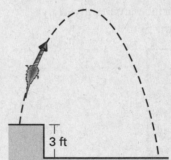

27. **Diving Board** A person steps off a 12-foot high diving board with 0 initial velocity. How many seconds does it take the person to hit the water?

28. **Diving Board** A person springs off a 12-foot high diving board with an initial velocity of 15 feet per second. How many seconds does it take the person to hit the water?

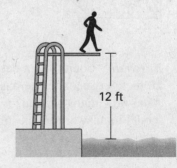

NAME _____ DATE _____

Practice C
For use with pages 533–538

Find the value of $b^2 - 4ac$ for the equation.

1. $15x^2 - 10x + 1 = 0$
2. $\frac{1}{3}x^2 - 6x + 3 = 0$
3. $4x^2 + x = \frac{1}{2}$
4. $-5x^2 - 12x = -8$
5. $6x^2 - 4 = -9x$
6. $8x - 12 = -3x^2$

Use the quadratic formula to solve the equation.

7. $-x^2 + x + 2 = 0$
8. $4x^2 - 6x = -2$
9. $3x^2 - 2 = -x$
10. $8x = -15x^2 - 1$
11. $8x^2 + 26x - 15 = 0$
12. $9x^2 - 9x = 1$
13. $-6x + 3x^2 = -1$
14. $7x - 5 = -4x^2$
15. $2x^2 = 6 + 19x$
16. $-9x^2 + 1 = 5x$
17. $-7 + 4x^2 = -6x$
18. $6x^2 + 16x = -1$

Find the x-intercepts of the graph of the equation.

19. $y = x^2 - 5x - 50$
20. $y = 2x^2 - 9x + 8$
21. $y = 4x^2 - 9x - 7$
22. $y = -2x^2 - 7x + 11$
23. $y = 5x^2 + 50x + 1$
24. $y = 2x^2 - 18x - 3$

25. **Surface Area**   The surface area of a rectangular box with a square base is 56 square inches. The surface area is given by $A = 2x^2 + 4xh$. Find x.

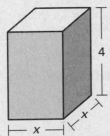

4

x

x

26. **Baton Twirling**   The baton twirler releases the baton into the air when it is 5 feet above the ground. The initial velocity of the baton is 30 feet per second. The twirler will catch the baton when it falls back to a height of 6 feet. How many seconds is the baton in the air?

27. **Cellular Phones**   For 1990 to 1997, the number of cellular phone subscribers s (in millions) in the United States can be approximated by the model

$$s = 0.84t^2 + 1.39t + 5.13$$

where $t = 0$ represents 1990. In which year did cellular phone companies have about 33.7 million subscribers?

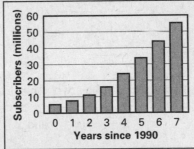

28. **Falling Glove**   You drop your softball glove out of your bedroom window, 28 feet above the ground. How long will it take for your glove to hit the ground?

29. Write a quadratic equation you would solve using the square root method. Write another quadratic equation you would solve using the quadratic formula. Solve each.

NAME _____ DATE _____

Reteaching with Practice

For use with pages 533–538

GOAL Use the quadratic formula to solve a quadratic equation and use quadratic models for real-life situations

> ### VOCABULARY
>
> The solutions of the quadratic equation $ax^2 + bx + c = 0$ are given by the **quadratic formula**
>
> $$x = \frac{-b \pm \sqrt{b^2 - 4ac}}{2a}$$ when $a \neq 0$ and $b^2 - 4ac \geq 0.$
>
> You can read this formula as "x equals the opposite of b, plus or minus the square root of b squared minus $4ac$, all divided by $2a$."

EXAMPLE 1 *Using the Quadratic Formula*

Solve $x^2 + 3x = 4$.

SOLUTION

You must rewrite the equation in standard form $ax^2 + bx + c = 0$ before using the quadratic formula.

$$x^2 + 3x = 4 \qquad \text{Write original equation.}$$

$$x^2 + 3x - 4 = 0 \qquad \text{Rewrite equation in standard form.}$$

$$1x^2 + 3x + (-4) = 0 \qquad \text{Identify } a = 1, b = 3, \text{ and } c = -4.$$

$$x = \frac{-3 \pm \sqrt{3^2 - 4(1)(-4)}}{2(1)} \qquad \begin{array}{l}\text{Substitute values into the quadratic} \\ \text{formula: } a = 1, b = 3, \text{ and } c = -4.\end{array}$$

$$x = \frac{-3 \pm \sqrt{9 + 16}}{2} \qquad \text{Simplify.}$$

$$x = \frac{-3 \pm \sqrt{25}}{2} \qquad \text{Simplify.}$$

$$x = \frac{-3 \pm 5}{2} \qquad \text{Solutions.}$$

The equation has two solutions:

$$x = \frac{-3 + 5}{2} = 1 \text{ and } x = \frac{-3 - 5}{2} = -4$$

Exercises for Example 1

Use the quadratic formula to solve the equation.

1. $x^2 - 4x + 3 = 0$ **2.** $x^2 + 9x + 20 = 0$ **3.** $x^2 + x = 6$

NAME _____ DATE _____

Reteaching with Practice

For use with pages 533–538

EXAMPLE 2 *Modeling Vertical Motion*
_____

You retrieve a football from a tree 25 feet above ground. You throw it downward with an initial speed of 20 feet per second. Use a vertical motion model to find how long it will take for the football to reach the ground.

SOLUTION

Because the football is thrown down, the initial velocity is $v = -20$ feet per second. The initial height is $s = 25$ feet. The football will reach the ground when the height is 0.

$h = -16t^2 + vt + s$	Choose the vertical motion model for a thrown object.
$h = -16t^2 + (-20)t + 25$	Substitute values for v and s into the vertical motion model.
$0 = -16t^2 - 20t + 25$	Substitute 0 for h.
$t = \dfrac{-(-20) \pm \sqrt{(-20)^2 - 4(-16)(25)}}{2(-16)}$	Substitute values into the quadratic formula: $a = -16$, $b = -20$, and $c = 25$.
$t = \dfrac{20 \pm \sqrt{2000}}{-32}$	Simplify.
$t \approx 0.773$ or -2.023	Solutions

The football will reach the ground about 0.773 seconds after it was thrown. As a solution, -2.023 doesn't make sense in the problem.

Exercise for Example 2
....

4. Rework Example 3 if the football is dropped from the tree with an initial speed of 0 feet per second.

NAME _____ DATE _____

Quick Catch-Up for Absent Students

For use with pages 533–539

The items checked below were covered in class on (date missed) _____

Lesson 9.5: Solving Quadratic Equations by the Quadratic Formula

_____ **Goal 1:** Use the quadratic formula to solve a quadratic equation. (pp. 533–534)

Material Covered:

_____ Example 1: Using the Quadratic Formula

_____ Example 2: Writing in Standard Form

_____ Student Help: Study Tip

_____ Example 3: Finding *x*-Intercepts of a Graph

Vocabulary:

quadratic formula, p. 533

_____ **Goal 2:** Use quadratic models for real-life situations. (p. 535)

Material Covered:

_____ Example 4: Modeling Vertical Motion

Activity 9.5: Writing a Program for the Quadratic Formula (p. 539)

_____ **Goal:** Program a graphing calculator or a computer to solve a quadratic equation.

_____ Student Help: Keystroke Help

_____ Other (specify) _____

_____

Homework and Additional Learning Support

_____ Textbook (specify) <u>pp. 536–538</u>_____

_____

_____ Internet: Extra Examples at www.mcdougallittel.com

_____ *Reteaching with Practice* worksheet (specify exercises)_____

_____ *Personal Student Tutor* for Lesson 9.5

NAME _____ DATE _____

Interdisciplinary Application

For use with pages 533–538

Current In An Electric Circuit

PHYSICS Electric current drives many modern devices, including laptops, electronic toys, and stereo equipment. Appliances operated by batteries use a form of electric current known as *direct current*. The battery supplies the power needed to use the appliance by providing a steady current, or flow of electrons, through the device.

Batteries have both a negative and a positive terminal. When these terminals are connected, a voltage V is created, which drives current I through the circuit. In the circuit shown here, a load is any device, such as a lamp, operated by electric current.

Some materials, called resistors, resist the flow of current, using the energy supplied for other than its intended purpose. Lamps, for example, give off heat in addition to light. The heat created is the unintended consequence of resistance. The total power supplied for light must account for this resistance and is given by the equation $P = VI - RI^2$. Power P is measured in watts, current I is measured in amperes, voltage V is measured in volts, and resistance R is measured in ohms.

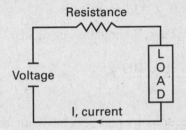

In Exercises 1–4, use the following information.

In a small circuit, the voltage is 110 volts, and the resistance of the circuit is 30 ohms. The load is expected to use 60 watts of power. Find the current.

1. Substitute the values for P, V, and R in the equation $P = VI - RI^2$.

2. Write the equation in standard form and give the values of a, b, and c.

3. Substitute the values of a, b, and c in the quadratic formula and simplify. Choose the most reasonable answer.

4. Check your result.

5. In a small circuit, the voltage is 220 volts and the resistance of the circuit is 37 ohms. The load is expected to use 175 watts of power. Find the current. Choose the most reasonable answer.

NAME _____ DATE _____

Challenge: Skills and Applications

For use with pages 533–538

In Exercises 1 and 2, the solution to a quadratic equation is given. Write an equation in standard form that has that solution.

Example: $x = \dfrac{-3 \pm \sqrt{297}}{16}$

Solution: The solution is in the form of the quadratic formula. Thus, to write one possible equation, you can find a, b, and c as follows.
$-b = -3$ so $b = 3$. Also, $2a = 16$, so $a = 8$

$b^2 - 4ac = 297$

$9 - 32c = 297 \Rightarrow -32c = 288 \Rightarrow c = -9$

Thus, an equation in standard from is $8x^2 + 3x - 9 = 0$

1. $x = \dfrac{7 \pm \sqrt{109}}{6}$

2. $x = \dfrac{-4.1 \pm \sqrt{20.17}}{-2.4}$

3. The graph of a parabola has x-intercepts at $\dfrac{3 \pm \sqrt{89}}{4}$. The axis of symmetry of any parabola is half-way between the x-intercepts. Find the equation of the axis of symmetry of the parabola.

In Exercises 4 and 5, use the following information.

The arc of a baseball thrown from the outfield is modeled by the equation

$$h = -16t^2 + 25t + 8$$

where h is the height (in feet) of the ball above ground t seconds after it is released.

4. Find all roots to the nearest tenth of a second.

5. How do you explain the physical fact that the ball will hit the ground exactly once?

TEACHER'S NAME _____ CLASS _____ ROOM _____ DATE _____

Lesson Plan

2-day lesson (See *Pacing the Chapter*, TE pages 500C–500D) | **For use with pages 540–547**

GOALS
1. **Use the discriminant to find the number of solutions of a quadratic equation.**
2. **Apply the discriminant to solve real-life problems.**

State/Local Objectives _____

_____

✓ Check the items you wish to use for this lesson.

STARTING OPTIONS
_____ Homework Check: TE page 536; Answer Transparencies
_____ Warm-Up or Daily Homework Quiz: TE pages 541 and 538, CRB page 83, or Transparencies

TEACHING OPTIONS
_____ Motivating the Lesson: TE page 542
_____ Concept Activity: SE page 540: CRB page 84 (Activity Support Master)
_____ Lesson Opener (Activity): CRB page 85 or Transparencies
_____ Examples: Day 1: 1–3, SE pages 541–542; Day 2: 4, SE page 543
_____ Extra Examples: Day 1: TE page 542 or Transp.; Day 2: TE page 543 or Transp.; Internet
_____ Closure Question: TE page 543
_____ Guided Practice: SE page 544; Day 1: Exs. 1–8; Day 2: none

APPLY/HOMEWORK
Homework Assignment
_____ Basic   Day 1: 9–26, 31; Day 2: 35–48; Quiz 2: 1–13
_____ Average   Day 1: 9–26, 31; Day 2: 35–48; Quiz 2: 1–13
_____ Advanced   Day 1: 9–26, 31–34; Day 2: 35–48; Quiz 2: 1–13

Reteaching the Lesson
_____ Practice Masters: CRB pages 86–88 (Level A, Level B, Level C)
_____ Reteaching with Practice: CRB pages 89–90 or Practice Workbook with Examples
_____ Personal Student Tutor

Extending the Lesson
_____ Applications (Real-Life): CRB page 92
_____ Math & History: SE page 547; CRB page 93; Internet
_____ Challenge: SE page 546; CRB page 94 or Internet

ASSESSMENT OPTIONS
_____ Checkpoint Exercises: Day 1: TE page 542 or Transp.; Day 2: TE page 543 or Transp.
_____ Daily Homework Quiz (9.6): TE page 546, CRB page 98, or Transparencies
_____ Standardized Test Practice: SE page 546; TE page 546; STP Workbook; Transparencies
_____ Quiz (9.4–9.6): SE page 547; CRB page 95

Notes _____

_____

_____

Algebra 1
Chapter 9 Resource Book

TEACHER'S NAME _____ CLASS _____ ROOM _____ DATE _____

Lesson Plan for Block Scheduling

1-day lesson (See *Pacing the Chapter,* TE pages 500C–500D)   For use with pages 540–547

GOALS 1. **Use the discriminant to find the number of solutions of a quadratic equation.**

2. **Apply the discriminant to solve real-life problems.**

State/Local Objectives _____

_____

_____

✓ **Check the items you wish to use for this lesson.**

STARTING OPTIONS

____ Homework Check: TE page 536; Answer Transparencies

____ Warm-Up or Daily Homework Quiz: TE pages 541 and
538, CRB page 83, or Transparencies

TEACHING OPTIONS

____ Motivating the Lesson: TE page 542

____ Concept Activity: SE page 540: CRB page 84 (Activity Support Master)

____ Lesson Opener (Activity): CRB page 85 or Transparencies

____ Examples 1–4: SE pages 541–543

____ Extra Examples: TE pages 542–543 or Transparencies; Internet

____ Closure Question: TE page 543

____ Guided Practice Exercises: SE page 544

APPLY/HOMEWORK

Homework Assignment

____ Block Schedule: 9–26, 31, 35–48; Quiz 2: 1–13

Reteaching the Lesson

____ Practice Masters: CRB pages 86–88 (Level A, Level B, Level C)

____ Reteaching with Practice: CRB pages 89–90 or Practice Workbook with Examples

____ Personal Student Tutor

Extending the Lesson

____ Applications (Real-Life): CRB page 92

____ Math & History: SE page 547; CRB page 93; Internet

____ Challenge: SE page 546; CRB page 94 or Internet

ASSESSMENT OPTIONS

____ Checkpoint Exercises: TE pages 542–543 or Transparencies

____ Daily Homework Quiz (9.6): TE page 546, CRB page 98, or Transparencies

____ Standardized Test Practice: SE page 546; TE page 546; STP Workbook; Transparencies

____ Quiz (9.4–9.6): SE page 547; CRB page 95

Notes _____

_____

_____

CHAPTER PACING GUIDE	
Day	**Lesson**
1	Assess Ch. 8; 9.1 (all)
2	9.2 (all); 9.3 (begin)
3	9.3 (end); 9.4 (all)
4	9.5 (all)
5	**9.6 (all)**
6	9.7 (all); 9.8 (begin)
7	9.8 (end); Review Ch. 9
8	Assess Ch. 9; 10.1 (all)

Algebra 1
Chapter 9 Resource Book

Available as
a transparency

NAME ——————————————— DATE ————

WARM-UP EXERCISES

For use before Lesson 9.6, pages 540–547

Evaluate the expression $b^2 - 4ac$ for the given values.

1. $a = 2, b = 8, c = 1$

2. $a = 3, b = 6, c = 3$

3. $a = 5, b = 3, c = 6$

4. $a = 2, b = 4, c = 0$

5. $a = 3, b = 0, c = -3$

DAILY HOMEWORK QUIZ

For use after Lesson 9.5, pages 533–539

1. Find the value of $b^2 - 4ac$ for $-2x^2 + 7x - 6$.

2. Use the quadratic formula to solve $6x^2 - x - 15 = 0$.

3. Write $-17x = -6x^2 - 12$ in standard form. Solve using the quadratic formula.

4. Find the x-intercepts of the graph of $y = 2x^2 - 5x + 1$.

5. Solve $9x^2 - 36 = 0$ by finding square roots or by using the quadratic formula.

6. The initial upward velocity of a golf ball is 60 ft/sec. How long will it take the ball to return to the ground?

NAME _____ DATE _____

Activity Support Master

For use with page 540

$x^2 + 3x - 2 = 0$

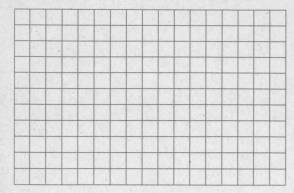

Number of x-intercepts: _____

Number of solutions to the equation: _____

$x^2 - 6x + 9 = 0$

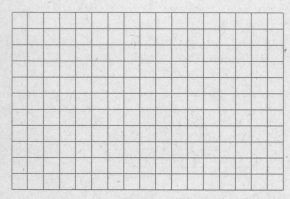

Number of x-intercepts: _____

Number of solutions to the equation: _____

$x^2 + 3 = 0$

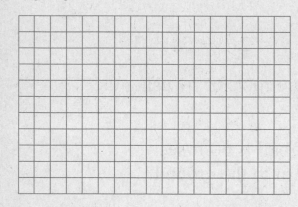

Number of x-intercepts: _____

Number of solutions to the equation: _____

_____

_____

_____

NAME _____ DATE _____

Practice A

For use with pages 541–547

...ch the discriminant with the graph.

... $b^2 - 4ac = 3$ **B.** $b^2 - 4ac = 0$ **C.** $b^2 - 4ac = -2$

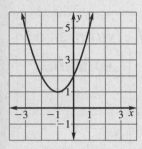

2.

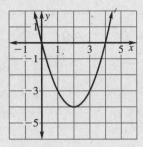

3.

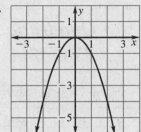

...ide how many solutions the equation has.

... $x^2 + 6x + 10 = 0$ **5.** $x^2 + 8x + 16 = 0$ **6.** $3x^2 - 5x + 2 = 0$

... $x^2 - 6x + 9 = 0$ **8.** $9x^2 - 6x + 1 = 0$ **9.** $15x^2 + 2x + 16 = 0$

... $x^2 + 4x - 12 = 0$ **11.** $-x^2 + 16x + 64 = 0$ **12.** $3x^2 + x - 1 = 0$

... $2x^2 + 7x + 50 = 0$ **14.** $2x^2 + 3x + 1 = 0$ **15.** $-3x^2 - 2x - 7 = 0$

...prints   In Exercises 16 and 17, use the following ...rmation.

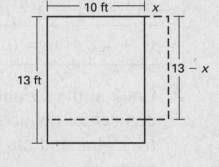

... want to build a shed in your backyard. You have blueprints which
... that the shed is 10 feet × 13 feet. You want to change the dimen-
... as shown at the right. The new area is modeled by the equation
$-x^2 + 3x + 130$.

What value of x, if any, will give an area of 140 ft²?

Is there any value of x for which the shed has an area of 124 ft²?
If so, what value of x?

Baseball   Suppose the equation $h = -\frac{1}{800}d^2 + \frac{1}{2}d + 5$ models the height
of a baseball at distance d after being hit from a height of 5 feet from the
ground. Which of the following heights are possible?

A. 57 feet   **B.** 100 feet

C. 75 feet   **D.** 50 feet

Painting   You are painting a house. While standing on a ladder that is
10 feet above the ground, you ask your friend to toss you a paintbrush.
The starting height of the paintbrush is 5 feet. Its starting velocity is
20 feet per second. Will the paintbrush reach you? Use the vertical
motion formula $h = -16t^2 + vt + s$ to answer the question.

Algebra 1
Chapter 9 Resource Book

Name _____ Date _____

Activity Lesson Opener

For use with pages 541–547

SET UP: Work with a partner.

1. Complete the table.

Quadratic Equation	Is $b^2 - 4ac$ >, =, or < 0?	Solution of Equation	Number of Solutions
$x^2 + 3x - 4 = 0$			
$2x^2 + x - 5 = 0$			
$-4x^2 + 2x + 1 = 0$			
$-x^2 - 4x - 4 = 0$			
$x^2 + 2x + 1 = 0$			
$3x^2 + x + 5 = 0$			
$-2x^2 + 2x - 1 = 0$			
$x^2 + 4x + 6 = 0$			

2. Look at the second and fourth columns of your table. Make a conjecture about the relationship between the value of $b^2 - 4ac$ and the number of solutions to a quadratic equation.

NAME _____ DATE _____

Practice B

For use with pages 541–547

Decide how many solutions the equation has.

1. $x^2 + 2x + 1 = 0$

2. $-2x^2 + 5x + 3 = 0$

3. $x^2 + 4x - 2 = 0$

4. $x^2 + 2x + 6 = 0$

5. $2x^2 + x - 15 = 0$

6. $-\frac{3}{2}x^2 + 2x + \frac{1}{2} = 0$

7. $x^2 - \frac{1}{2}x + 8 = 0$

8. $\frac{4}{3}x^2 + 4x + 3 = 0$

9. $-25x^2 + 10x - 1 = 0$

10. $-12x^2 + 19x - 5 = 0$

11. $6x^2 + 25x + 21 = 0$

12. $3x^2 - 5x + 4 = 0$

13. **Geometry** Is it possible for the rectangle below to have a perimeter of 32 inches and an area of 70 square inches?

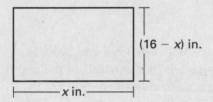

14. **Geometry** Is it possible for the rectangle below to have a perimeter of 40 inches and an area of 75 square inches?

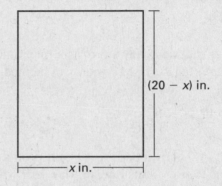

15. **Profit** Your company's profit (in thousands of dollars) for the last 10 years can be modeled by

$$P = 7.54t^2 - 2.23t + 8.12$$

where t represents the number of years since 1990. If profits continue to rise at this rate, during what year will the company have a profit of $1,070,000?

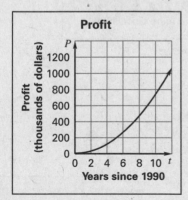

Football **In Exercises 16–18, use the following information.**

You kick a football with an initial velocity of 40 feet per second. The equation $h = -16t^2 + 40t$ models the height the football will reach t seconds after it is kicked.

16. Is it possible for the ball to reach 16 feet?

17. Is it possible for the ball to reach 25 feet?

18. Is it possible for the ball to reach 30 feet?

Tennis **In Exercises 19 and 20, use the vertical motion model $h = -16t^2 + vt + s$ and the following information.**

You and your friend are walking around a tennis court that has a 10-foot high fence around it. You see two tennis balls lying outside the fence.

19. You pick up a ball and try to throw it from a height of 4.5 feet over the fence. You throw it with an initial velocity of 22 feet per second. Did the ball make it over the fence?

20. Your friend throws the other ball from a height of 5 feet with an initial velocity of 17 feet per second. Did the ball make it over the fence?

Algebra 1
Chapter 9 Resource Book

Practice C

For use with pages 541–547

Decide how many solutions the equation has.

1. $x^2 + 12x + 32 = 0$ **2.** $3x^2 + 5x - 2 = 0$ **3.** $5x^2 - 8x + 9 = 0$

4. $2x^2 - 12x + 18 = 0$ **5.** $-2x^2 - 5x - 6 = 0$ **6.** $x^2 - 2x + \frac{1}{3} = 0$

7. $-\frac{7}{2}x^2 - 4x - 5 = 0$ **8.** $\frac{3}{2}x^2 - 3x + \frac{3}{2} = 0$ **9.** $-5x^2 - x + 1 = 0$

10. $-2x^2 + 3x - 1 = 0$ **11.** $18x^2 - 11x + 2 = 0$ **12.** $-3x^2 + 12x - 12 = 0$

Find the value of *c* such that the equation will have two solutions, one solution, and no real solution. Then sketch the graph of the equation for each value of *c* that you chose.

13. $x^2 + 6x + c = 0$ **14.** $2x^2 - 4x + c = 0$ **15.** $3x^2 - 6x + c = 0$

16. *Profit*   Your company's profit (in thousands of dollars) for the last 12 years can be modeled by

$$P = \frac{10}{3}t^2 + \frac{50}{3}t + 100$$

where $t = 0$ corresponds to 1985. If profits continue to rise at this rate, in what year will the company have a profit of $1,100,000?

17. *Revenue*   Between 1990 and 2000, the revenue earned (in thousands of dollars) from exercise equipment can be modeled by

$$R = \frac{1}{13}t^2 - \frac{11}{65}t + 10$$

where $t = 0$ corresponds to 1990. If the trend continues, in what year will revenue reach $37,000?

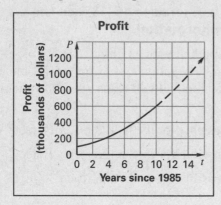

Making a Basket   **In Exercises 18 and 19, use the following information.**

Patty is only 7 years old, but she insists that she can play basketball with a regulation-height hoop. She can throw the ball with an initial velocity of 20 feet per second.

18. Patty needs to be able to throw the ball 7 feet straight up. Can she make the basket?

19. If Patty's dad gets her a 1-foot tall box to stand on, can Patty make a basket?

NAME _____ DATE _____

Reteaching with Practice

For use with pages 541–547

GOAL **Use the discriminant to find the number of solutions of a quadratic equation and apply the discriminant to solve real-life problems**

VOCABULARY

The **discriminant** is the expression inside the radical in the quadratic formula, $b^2 - 4ac$.

Consider the quadratic equation $ax^2 + bx + c = 0$.

- If $b^2 - 4ac$ is positive, then the equation has two solutions.
- If $b^2 - 4ac$ is zero, then the equation has one solution.
- If $b^2 - 4ac$ is negative, then the equation has no real solution.

EXAMPLE 1 *Finding the Number of Solutions*

Find the value of the discriminant and use the value to tell if the equation has *two solutions, one solution,* or *no real solution.*

a. $3x^2 - 2x - 1 = 0$ **b.** $x^2 - 8x + 16 = 0$ **c.** $x^2 - 4x + 5 = 0$

SOLUTION

a. $b^2 - 4ac = (-2)^2 - 4(3)(-1)$ Substitute 3 for a, -2 for b, -1 for c.

$\qquad\qquad\quad = 4 + 12$ Simplify.

$\qquad\qquad\quad = 16$ Discriminant is positive.

The discriminant is positive, so the equation has two solutions.

b. $b^2 - 4ac = (-8)^2 - 4(1)(16)$ Substitute 1 for a, -8 for b, 16 for c.

$\qquad\qquad\quad = 64 - 64$ Simplify.

$\qquad\qquad\quad = 0$ Discriminant is zero.

The discriminant is zero, so the equation has one solution.

c. $b^2 - 4ac = (-4)^2 - 4(1)(5)$ Substitute 1 for a, -4 for b, 5 for c.

$\qquad\qquad\quad = 16 - 20$ Simplify.

$\qquad\qquad\quad = -4$ Discriminant is negative.

The discriminant is negative, so the equation has no real solution.

Exercises for Example 1

Tell if the equation has *two solutions, one solution,* or *no real solution.*

1. $x^2 - 10x + 25 = 0$ **2.** $2x^2 - x - 1 = 0$ **3.** $x^2 + 2x + 4 = 0$

4. $-x^2 + 6x - 9 = 0$ **5.** $-2x^2 - 5x - 4 = 0$ **6.** $3x^2 + 2x - 16 = 0$

NAME _____ DATE _____

Reteaching with Practice

For use with pages 541–547

EXAMPLE 2 *Using the Discriminant in a Real-Life Problem*

You work as an accountant for a sporting goods company. You have been asked to project the revenue of the company. The revenue of the company from 1990 to 1995 can be modeled by

$$R = 1.23t^2 - 2.22t + 8.5$$

where R is the revenue in millions of dollars and t is the number of years since 1990. Use the model to predict whether the revenue will reach 90 million dollars.

SOLUTION

Set the revenue model equal to 90 and use the discriminant to determine the number of solutions of the quadratic revenue model.

$R = 1.23t^2 - 2.22t + 8.5$	Write revenue model.
$90 = 1.23t^2 - 2.22t + 8.5$	Substitute 90 for R.
$0 = 1.23t^2 - 2.22t - 81.5$	Rewrite equation in standard form.
$0 = 1.23t^2 + (-2.22)t + (-81.5)$	Identify $a = 1.23$, $b = -2.22$, and $c = -81.5$.
$b^2 - 4ac = (-2.22)^2 - 4(1.23)(-81.5)$	Substitute 1.23 for a, -2.22 for b, -81.5 for c.
$= 4.9284 + 400.98$	Simplify.
$= 405.9084$	Discriminant is positive.

The discriminant is positive, so the equation has two solutions. You predict that the company's revenue will reach 90 million dollars.

Exercises for Example 2

7. Use the discriminant to show that the revenue for the company will reach $150 million.

8. Use a graphing calculator to find how many years it will take for the revenue to reach $90 million.

Algebra 1
Chapter 9 Resource Book

NAME _____ DATE _____

Quick Catch-Up for Absent Students

For use with pages 540–547

The items checked below were covered in class on (date missed) _____

Activity 9.6: Investigating Applications of the Discriminant (p. 540)

_____ **Goal:** Determine the number of solutions of a quadratic equation.

_____ Student Help: Look Back

Lesson 9.6: Applications of the Discriminant

_____ **Goal 1:** Use the discriminant to find the number of solutions of a quadratic equation. (pp. 541–542)

Material Covered:

_____ Example 1: Finding the Number of Solutions

_____ Example 2: Finding the Number of x-Intercepts

_____ Student Help: Look Back

_____ Example 3: Changing the Value of c

Vocabulary:

discriminant, p. 541

_____ **Goal 2:** Apply the discriminant to solve real-life problems. (p. 543)

Material Covered:

_____ Example 4: Using the Discriminant

_____ Other (specify) _____

_____

Homework and Additional Learning Support

_____ Textbook (specify) pp. 544–547 _____

_____

_____ Internet: Extra Examples at www.mcdougallittel.com

_____ *Reteaching with Practice* worksheet (specify exercises)_____

_____ *Personal Student Tutor* for Lesson 9.6

Real-Life Application:
When Will I Ever Use This?

For use with pages 541–547

Factory Sales

You and your friend want to start a small business dealing with pagers. You want to know if this is a reasonable business or a risky business adventure. To find this out, you decide to find the factory sales (in millions of dollars) of pagers from 1990 to 1996. In 1990 there were $118 million in factory sales and in 1996 there were $370 million in factory sales.

You decide that for the business to prosper there should be about $500,000,000 in factory sales of pagers. The equation $S = 5.4t^2 + 8.3t + 122.5$ models the pager factory sales, where S is the sales (in millions of dollars) and t is the year with $t = 0$ representing 1990.

In Exercises 1-3, use the information above.

1. Using the equation $S = 5.4t^2 + 8.3t + 122.5$, the number $500,000,000 in factory sales would correspond to an S-value of 500. Why?

2. Substitute the S-value in Exercise 1 in the equation and rewrite the equation in standard form.

3. Use the discriminant to decide if the equation you found in Exercise 2 has a solution. Will this business be risky or not?

In Exercises 4 and 5, use the following information.

You and your friend are also considering a business repairing laserdisc players. The factory sales of laserdisc players from 1990 to 1996 can be modeled by the equation $S = -5.3t^2 + 34.25t + 61.4$ where S is the sales (in millions of dollars) and t is the year with $t = 0$ representing 1990.

4. You decide that for the business to be successful there has to be about $120,000,000 in factory sales of laserdisc players. Substitute the corresponding S-value in the equation and rewrite the equation in standard form.

5. Decide whether the business will be successful.

NAME _____ DATE _____

Math and History Application

For use with page 547

HISTORY   The earliest known solutions of quadratic equations are found in Babylonian textbooks from around 2000 B.C. The method of finding positive real roots is exactly the same as the method taught today, except Babylonians used words where we would use symbols. The words the Babylonians used showed the connection between math theory and practical application. For example, an unknown x was often called "the side" and its square "the square."

MATH   A problem in a Babylonian math book may have been written as, "I have subtracted the side of the square from the area, and the result is 870." We would write this as the quadratic equation $x^2 - x = 870$. The Babylonians would have used the following steps to solve the equation:

Take 1, *the coefficient of the linear term.*

Take one-half of 1, *result* $\frac{1}{2}$.

Multiply $\frac{1}{2}$ by $\frac{1}{2}$, *result* $\frac{1}{4}$.

Add $\frac{1}{4}$ to 870, *result* $870\frac{1}{4}$.

Take the square root of this, *which is* $29\frac{1}{2}$.

Take the $\frac{1}{2}$ *which you have multiplied by itself, and add it to* $29\frac{1}{2}$.

Result 30: *this is the required side of the square.*

This algorithm is also known as "completing the square." Babylonian mathematicians solved equations using rules explained as operations.

1. Use the discriminant to find the number of solutions of the equation. How many reasonable solutions are there? Explain your answer.

2. Use the quadratic formula to solve the equation.

3. Discuss why the Babylonians used words instead of letters in algebra. Using the example of a Babylonian equation above, write a quadratic equation using words and then in symbols. Which version is easier to understand? Explain your answer.

Challenge: Skills and Applications

For use with pages 541–547

In Exercises 1–6, find the value(s) of k for which the equation has exactly one real root.

1. $x^2 - 4kx + 36 = 0$

2. $x^2 + 10x + 3k = 0$

3. $4x^2 - 5kx + 81 = 0$

4. $2x^2 + 5kx + \frac{25}{2} = 0$

5. $\frac{1}{3}x^2 - 4x - 75k = 0$

6. $3kx^2 + 2x + 3k = 0$

In Exercises 7–10, find the values of k for which the equation has two real roots. Write your answer as an inequality.

7. $x^2 - 6x - 2k = 0$

8. $x^2 + 5x + 4k = 0$

9. $2x^2 + 2x - 5k = 0$

10. $3x^2 + 7x + 5k = 0$

In Exercises 11–14,

    a. Find the discriminant of the equation in terms of p, q, or both.

    b. Suppose $q > p > 0$. Tell how many real roots the equation has.

11. $x^2 + 2qx + p^2 = 0$

12. $px^2 + qx - p = 0$

13. $qx^2 + px + q = 0$

14. $px^2 - 2px + p = 0$

NAME _____ DATE _____

Quiz 2

For use after Lessons 9.4–9.6

1. Solve the equation algebraically. Check the solutions graphically. *(Lesson 9.4)*

$$\frac{1}{2}x^2 = 50$$

2. Use the quadratic formula to solve $x^2 - x = 2$. *(Lesson 9.5)*

3. Find the *x*-intercepts of the graph of the equation $x^2 - 3x - 18 = 0$. *(Lesson 9.5)*

4. Decide how many solutions the equation $x^2 - 4x + 4 = 0$ has. *(Lesson 9.6)*

5. Find the value of c so that $x^2 + 2x + c = 0$ will have two solutions. *(Lesson 9.6)*

6. Use a vertical motion model to find how long it will take a walnut to reach the ground if it falls from the top of a 55-foot tree. *(Lesson 9.5)*

Answers

1. _____
2. _____
3. _____
4. _____
5. _____
6. _____

Lesson 9.6

Lesson Plan

1-day lesson (See *Pacing the Chapter,* TE pages 500C–500D)   For use with pages 548–553

GOALS
1. **Sketch the graph of a quadratic inequality.**
2. **Use quadratic inequalities as real-life models.**

State/Local Objectives _____

_____

✓ **Check the items you wish to use for this lesson.**

STARTING OPTIONS
____ Homework Check: TE page 544; Answer Transparencies
____ Warm-Up or Daily Homework Quiz: TE pages 548 and 546, CRB page 98, or Transparencies

TEACHING OPTIONS
____ Lesson Opener (Application): CRB page 99 or Transparencies
____ Graphing Calculator Activity with Keystrokes: CRB page 100
____ Examples 1–4: SE pages 548–550
____ Extra Examples: TE pages 549–550 or Transparencies
____ Closure Question: TE page 550
____ Guided Practice Exercises: SE page 551

APPLY/HOMEWORK
Homework Assignment
____ Basic   13–33, 39, 40, 44–66 even
____ Average   13–33, 39, 40, 44–66 even
____ Advanced   13–33, 39, 40–43, 44–66 even

Reteaching the Lesson
____ Practice Masters: CRB pages 101–103 (Level A, Level B, Level C)
____ Reteaching with Practice: CRB pages 104–105 or Practice Workbook with Examples
____ Personal Student Tutor

Extending the Lesson
____ Applications (Real-Life): CRB page 107
____ Challenge: SE page 553; CRB page 108 or Internet

ASSESSMENT OPTIONS
____ Checkpoint Exercises: TE pages 549–550 or Transparencies
____ Daily Homework Quiz (9.7): TE page 553, CRB page 111, or Transparencies
____ Standardized Test Practice: SE page 553; TE page 553; STP Workbook; Transparencies

Notes _____

_____

_____

TEACHER'S NAME _____ CLASS _____ ROOM _____ DATE _____

Lesson Plan for Block Scheduling

Half-day lesson (See *Pacing the Chapter,* TE pages 500C–500D)   **For use with pages 548–553**

GOALS
1. **Sketch the graph of a quadratic inequality.**
2. **Use quadratic inequalities as real-life models.**

State/Local Objectives _____

_____

_____

✓ **Check the items you wish to use for this lesson.**

STARTING OPTIONS

____ Homework Check: TE page 544; Answer Transparencies

____ Warm-Up or Daily Homework Quiz: TE pages 548 and
       546, CRB page 98, or Transparencies

TEACHING OPTIONS

____ Lesson Opener (Application): CRB page 99 or Transparencies

____ Graphing Calculator Activity with Keystrokes: CRB page 100

____ Examples 1–4: SE pages 548–550

____ Extra Examples: TE pages 549–550 or Transparencies

____ Closure Question: TE page 550

____ Guided Practice Exercises: SE page 551

APPLY/HOMEWORK

Homework Assignment (See also the assignment for Lesson 9.8.)

____ Block Schedule: 13–33, 39, 40, 44–66 even

Reteaching the Lesson

____ Practice Masters: CRB pages 101–103 (Level A, Level B, Level C)

____ Reteaching with Practice: CRB pages 104–105 or Practice Workbook with Examples

____ Personal Student Tutor

Extending the Lesson

____ Applications (Real-Life): CRB page 107

____ Challenge: SE page 553; CRB page 108 or Internet

ASSESSMENT OPTIONS

____ Checkpoint Exercises: TE pages 549–550 or Transparencies

____ Daily Homework Quiz (9.7): TE page 553, CRB page 111, or Transparencies

____ Standardized Test Practice: SE page 553; TE page 553; STP Workbook; Transparencies

Notes _____

_____

_____

CHAPTER PACING GUIDE	
Day	Lesson
1	Assess Ch. 8; 9.1 (all)
2	9.2 (all); 9.3 (begin)
3	9.3 (end); 9.4 (all)
4	9.5 (all)
5	9.6 (all)
6	**9.7 (all)**; 9.8 (begin)
7	9.8 (end); Review Ch. 9
8	Assess Ch. 9; 10.1 (all)

Lesson 9.7

NAME _____ DATE _____

WARM-UP EXERCISES

For use before Lesson 9.7, pages 548–553

Tell whether each point lies on the graph of $y = 3x + 2$.

1. $(0, 2)$

2. $(-1, 4)$

3. $(-2, -1)$

4. $(3, 11)$

5. $(-1, -1)$

DAILY HOMEWORK QUIZ

For use after Lesson 9.6, pages 540–547

Tell if the equation has *two solutions, one solution,* or *no real solution.*

1. $4x^2 - 8x + 5 = 0$

2. $-0.5x^2 + 11x - 33 = 0$

3. Evaluate the discriminant of $-\dfrac{3}{4}x^2 - \dfrac{5}{2}x - 2$. Does the graph of the related function cross the x-axis?

4. Find the values of c so that $-x^2 + 4x + c$ will have *two solutions, one solution,* and *no real solution.*

5. An antelope must jump 5.5 ft to clear a fence. It can jump with an intitial vertical velocity of 20 ft/sec. Using the discriminant and the vertical motion model $h = -16t^2 + vt + s$, can the antelope make the jump?

Algebra 1
Chapter 9 Resource Book

NAME _____ DATE _____

Application Lesson Opener

For use with pages 548–553

1. You are fencing in a square dog run. The minimum area of the dog run must be 36 ft². If x represents the length of one side of the dog run, which inequality describes the situation? Why?

   A. $x^2 \leq 36$

   B. $x^2 > 36$

   C. $x^2 < 36$

   D. $x^2 \geq 36$

2. You have a patio in your backyard that you want to enlarge. Your existing patio has dimensions of 5 ft by 6 ft. You want to increase each dimension of the patio by the same amount. You want the area to be greater than 35 ft². If x represents the amount by which you increase each dimension, which inequality describes the situation? Why?

   A. $(x + 5)(x + 6) > 35$

   B. $(5 - x)(6 - x) > 35$

   C. $(x + 5)(x + 6) < 35$

   D. $(x - 5)(x - 6) < 35$

3. You are buying an area rug for a room. The dimensions of the room are 10 ft by 12 ft. You want to leave an equal width of floor uncovered along each side of the rug. You want the rug to have a maximum area of 99 ft². If x represents the width of the floor that will be uncovered on each side, which inequality describes the situation? Why?

   A. $(2x - 10)(2x - 12) > 99$

   B. $(10 - 2x)(12 - 2x) \leq 99$

   C. $(x + 10)(x + 12) \leq 99$

   D. $(x + 10)(x + 12) < 99$

Graphing Calculator Activity Keystrokes

For use with page 553.

Keystrokes for Exercises 41–43

TI-82

[Y=] 0.09 [X,T,θ] [x^2] [−] 5.29 [X,T,θ] [+]

114.69 [ENTER]

[(-)] 0.23 [X,T,θ] [x^2] [+] 2.40 [X,T,θ] [+]

71.39 [ENTER]

[WINDOW] [ENTER]

0 [ENTER]

20 [ENTER]

1 [ENTER]

50 [ENTER]

120 [ENTER]

5 [ENTER]

[2nd] [DRAW] 7

[2nd] [Y-VARS] 1 1 [,]

[2nd] [Y-VARS] 1 2 [)] [ENTER]

[TRACE]

TI-83

[Y=] 0.09 [X,T,θ,n] [x^2] [−] 5.29 [X,T,θ,n] [+]

114.69 [ENTER]

[(-)] 0.23 [X,T,θ,n] [x^2] [+] 2.40 [X,T,θ,n] [+]

71.39 [ENTER]

[WINDOW]

0 [ENTER]

20 [ENTER]

1 [ENTER]

50 [ENTER]

120 [ENTER]

5 [ENTER]

[2nd] [DRAW] 7

[2nd] [Y-VARS] 1 1 [,]

[2nd] [Y-VARS] 1 2 [)] [ENTER]

[TRACE]

SHARP EL-9600c

[Y=] 0.09 [X/θ/T/n] [x^2] [−] 5.29 [X/θ/T/n] [+]

114.69 [ENTER]

[(-)] 0.23 [X/θ/T/n] [x^2] [+] 2.40 [X/θ/T/n] [+]

71.39 [ENTER]

[WINDOW]

0 [ENTER]

20 [ENTER]

1 [ENTER]

50 [ENTER]

120 [ENTER]

5 [ENTER]

[2ndF] [DRAW][G] 1 [−] [▶] [−] [−]

[GRAPH]

[TRACE]

CASIO CFX-9850Ga PLUS

From the main menu, choose GRAPH.

[F3] [F6] [F1] 0.09 [X,θ,T] [x^2] [−] 5.29 [X,θ,T]

[+] 114.69 [EXE]

[F3] [F6] [F2] [(-)] 0.23 [X,θ,T] [x^2] [+]

2.40 [X,θ,T] [+] 71.39 [EXE]

[SHIFT] [F3]

0 [EXE]

20 [EXE]

1 [EXE]

50 [EXE]

120 [EXE]

5 [EXE]

[EXIT]

[F6]

[F1]

Practice A

For use with pages 548–553

Decide whether the ordered pair is a solution of the inequality.

1. $y > x^2 + 6x$, $(1, 7)$

2. $y < 3x^2 - 8x$, $(2, -1)$

3. $y \le x^2 - 5x - 6$, $(0, -10)$

4. $y \ge -x^2 + 2x - 3$, $(-3, 0)$

5. $y > -8x^2 + 4x - 6$, $(-1, -2)$

6. $y \le -2x^2 - 11x - 4$, $(-4, 9)$

7. $y < 2x^2 - 3x + 4$, $(-1, -8)$

8. $y \ge 2x^2 + 5x + 3$, $(3, 12)$

9. $y \le 4x^2 + 3x + 8$, $(4, 82)$

Match the graph with its inequality.

A. $y > -x^2 + 4x + 2$

B. $y \ge \frac{1}{2}x^2 - 5$

C. $y < -x^2 - 4x - 3$

D. $y \ge -x^2 - 2$

E. $y \ge x^2 - 3x + 1$

F. $y < 3x^2 + 9x - 1$

10.
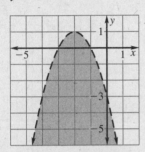

11.
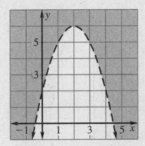

12.
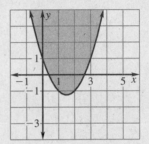

13.
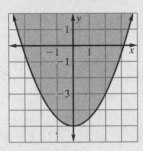

14.
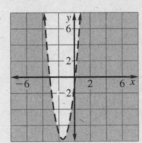

15.
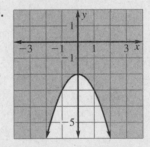

Sketch the graph of the inequality.

16. $y \le x^2 + 4$

17. $y > -x^2 - 2$

18. $y \ge -x^2 + 6$

19. $y < -2x^2 + 8$

20. $y \le 3x^2 + 5x$

21. $y \le x^2 + 5x + 6$

22. $y > x^2 - x - 12$

23. $y \ge -x^2 + 3x - 4$

24. $y < 2x^2 - 5x + 3$

25. *Water Use* For 1940 to 1980, the daily water use W (in billions of gallons) in the United States followed the quadratic model

$$W = 0.050t^2 + 6.109t + 130.727$$

where $t = 0$ represents 1940. Suppose this model had been used to predict the water use for 1980 through 1995. The actual use in 1995 was 402 billion gallons. Was this less or more than the predicted consumption for 1995? Explain.

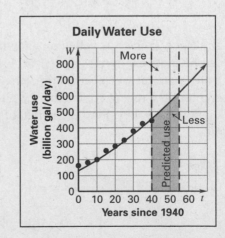

NAME _____ DATE _____

Practice B

For use with pages 548–553

Decide whether the ordered pair is a solution of the inequality.

1. $y < 3x^2 - 8x, (2, -2)$

2. $y > x^2 + 6x - 3, (1, 4)$

3. $y \leq 3x^2 - 5x - 6, (0, -8)$

4. $y \geq -x^2 + 2x - 3, (-3, -12)$

5. $y > -8x^2 + 4x - 6, (-1, -6)$

6. $y \leq -2x^2 - 11x + 4, (-5, 9)$

Match the graph with its inequality.

A. $y > -x^2 + 4x + 3$

B. $y \geq \frac{1}{2}x^2 + 4x$

C. $y < -x^2 - 4x + 3$

D. $y \leq -\frac{1}{2}x^2 + 3$

E. $y \geq 3x^2 - 4x + 1$

F. $y < 3x^2 + 4x - 1$

7.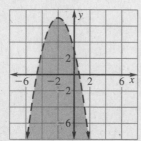

8.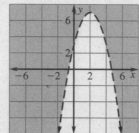

9.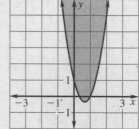

10.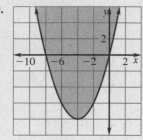

11.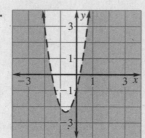

12.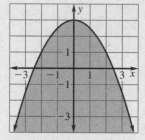

Sketch a graph of the inequality.

13. $y > x^2 - 5x$

14. $y \leq 2x^2 + 6x$

15. $y \geq -x^2 + 4x$

16. $y \leq -x^2 + 6x + 4$

17. $y > -x^2 + 5x + 6$

18. $y \geq -x^2 - 7x + 10$

19. $y \leq 2x^2 + 2x + 1$

20. $y > -3x^2 + 2x - 1$

21. $y > 2x^2 + 3x - 2$

22. $y < -2x^2 + 16x + 1$

23. $y \leq 4x^2 - 8x + 3$

24. $y \geq -3x^2 + 12x - 7$

25. **Toy Race Track**   A valley on a toy race track can be modeled by $y \leq \frac{1}{50}x^2 - \frac{4}{5}x + 10$, where x and y are measured in centimeters. Sketch the graph of the region containing the pillars that support the valley. Find the height of the tallest pillar.

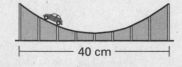

|← 40 cm →|

26. **Wrapping**   You have a rectangular piece of wrapping paper whose perimeter is at most 20 inches. You want to use it to cover a cubical box of length 2 inches. Sketch the graph that shows the possible area of the paper. Is there a possibility that you could cover the box?

NAME _____ DATE _____

Practice C
For use with pages 548–553

Decide whether the ordered pair is a solution of the inequality.

1. $y < 3x^2 - 8x, (2, -3)$

2. $y > x^2 + 6x - 8, (1, -1)$

3. $y \leq 3x^2 - 5x - 6, (0, -10)$

4. $y \geq -x^2 + 2x - 3, (-3, -18)$

5. $y > -8x^2 + 4x - 6, (-1, -17)$

6. $y \leq -2x^2 - 11x + 4, (-5, 10)$

Match the graph with its inequality.

A. $y > -\frac{1}{2}x^2 + 4x + 2$

B. $y \geq \frac{1}{2}x^2 - 4x$

C. $y \leq -\frac{1}{2}x^2 - 4x - 2$

7.

8.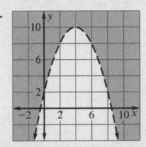

9.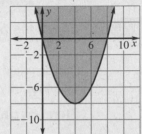

Sketch the graph of the inequality.

10. $y > x^2 - 6x$

11. $y \geq -2x^2 + 5x$

12. $y \leq x^2 + 4x + 2$

13. $y < -x^2 + 6x - 2$

14. $y \geq -2x^2 + 4x + 8$

15. $y < 3x^2 - 18x + 2$

16. $y > 3x^2 + 12x$

17. $y \geq 5x^2 + 10x + 7$

18. $y > 3x^2 - 12x + 12$

19. $y < -3x^2 + 6x + 7$

20. $y > -2x^2 - 8x + 2$

21. $y \leq 4x^2 - x + 5$

Sidewalk **In Exercises 22–24, use the following information.**

A sidewalk of uniform width x is to be built on two sides of a corner lot, as shown at the right. The lot was originally 60 feet by 90 feet.

22. Write an expression in terms of x for the length and width of the part of the remaining lot.

23. Write an expression for the area of the remaining lot.

24. The city agrees to leave at least 80% of the area of the original lot after the sidewalk is built. Use a graphing calculator to graph the possible area of the remaining lot.

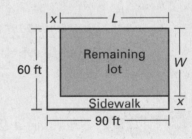

25. *Mounting a Picture* A rectangular picture is twice as long as it is wide and is to be mounted on a mat with the same proportion. Sketch the graph that shows the minimum area of a mat needed to hold a picture of a given width. If the picture is 10 inches wide, can it be mounted on a mat of 110 square inches? Of 210 square inches?

26. *Bridge* The vertical supports on an arch on a bridge can be modeled by

$$y \leq -\frac{1}{500}x^2 + \frac{2}{5}x$$

where x and y are measured in meters. Sketch a graph of one arch of the bridge and indicate the possible heights of the supports.

Reteaching with Practice

For use with pages 548–553

GOAL **Sketch the graph of a quadratic inequality**

VOCABULARY

The following are types of **quadratic inequalities.**

$y < ax^2 + bx + c$ $y \le ax^2 + bx + c$

$y > ax^2 + bx + c$ $y \ge ax^2 + bx + c$

The **graph** of a quadratic inequality consists of the graph of all ordered pairs (x, y) that are solutions of the inequality.

EXAMPLE 1 *Checking Solutions*

Decide whether the ordered pairs $(-4, -5)$ and $(0, 2)$ are solutions of the inequality $y < x^2 + 5x$.

SOLUTION

$y < x^2 + 5x$.   Write original inequality.

$-5 \overset{?}{<} (-4)^2 + 5(-4)$ Substitute -4 for x and -5 for y.

$-5 < -4$ True.

Because $-5 < -4$, the ordered pair $(-4, -5)$ is a solution of the inequality.

$y < x^2 + 5x$ Write original inequality.

$2 \overset{?}{<} (0)^2 + 5(0)$ Substitute 0 for x and 2 for y.

$2 \not< 0$ False.

Because $2 \not< 0$, the ordered pair $(0, 2)$ is not a solution of the inequality.

Exercises for Example 1

Decide whether the ordered pair is a solution of the inequality.

1. $y \ge x^2 - 2x$, $(2, 0)$ **2.** $y < 2x^2 + x$, $(1, -1)$ **3.** $y > x^2 - 3x$, $(2, -3)$

EXAMPLE 2 *Graphing a Quadratic Inequality*

Sketch the graph of $y \ge 2x^2 + 6x$.

SOLUTION

Sketch the parabola $y = 2x^2 + 6x$ using a solid line because the inequality is $\ge$. The parabola opens up. Test the point $(2, 0)$ which is not on the parabola.

$y \ge 2x^2 + 6x$ Write original inequality.

$0 \overset{?}{\ge} 2(2)^2 + 6(2)$ Substitute 2 for x and 0 for y.

$0 \not\ge 20$ False.

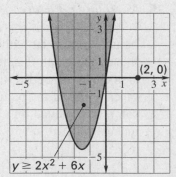

Reteaching with Practice

For use with pages 548–553

Because 0 is not greater than or equal to 20, the ordered pair $(2, 0)$ is not a solution.

The point $(2, 0)$ is not a solution and it is outside the parabola, so the graph of $y \geq 2x^2 + 6x$ is all the points inside or on the parabola.

Exercises for Example 2

Sketch the graph of the inequality.

4. $y \leq x^2 + 4x + 4$ **5.** $y \geq 3x^2 - 12$ **6.** $y \leq x^2 - 10x + 9$

EXAMPLE 3   *Using a Quadratic Inequality Model*

A rectangular pen is 4 feet longer than it is wide. The area of the pen is more than 32 square feet. Sketch the inequality that describes the possible dimensions of the pen.

SOLUTION

Let x represent the width of the pen. So, the pen's length is $x + 4$ and the area of the pen is $x(x + 4)$. Since the area of the pen is greater than 32 square feet, you have

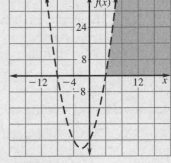

$$x(x + 4) > 32$$

$$x^2 + 4x > 32$$

$$x^2 + 4x - 32 > 0.$$

Sketch the parabola using a dashed line. The parabola opens up. Test the point $(0, 0)$ to determine what portion of the graph to shade. Since the point $(0, 0)$ does not satisfy the inequality, the graph of $x^2 + 4x - 32 > 0$ is all the points outside the parabola. Since the length cannot be negative, only points in the first quadrant are considered. So, the width should be greater than 4 feet and the length greater than 8 feet.

Exercises for Example 3

7. Give two possible scenarios for the dimensions of the rectangular pen in Example 3.

8. Suppose a rectangular pen is 10 feet longer than it is wide and its area is more than 96 square feet. Sketch the inequality that describes the dimensions of the rectangular pen.

Quick Catch-Up for Absent Students

For use with pages 548–553

The items checked below were covered in class on (date missed) _____

Lesson 9.7: Graphing Quadratic Inequalities

____ **Goal 1:** Sketch the graph of a quadratic inequality. (pp. 548–549)

Material Covered:

____ Example 1: Checking Points

____ Student Help: Look Back

____ Example 2: Graphing a Quadratic Inequality

____ Example 3: Graphing a Quadratic Inequality

Vocabulary:

quadratic inequalities, p. 548   graph of a quadratic inequality, p. 548

____ **Goal 2:** Use quadratic inequalities as real-life models. (p. 550)

Material Covered:

____ Example 4: Using a Quadratic Inequality Model

____ Other (specify) _____

_____

Homework and Additional Learning Support

____ Textbook (specify) <u>pp. 551–553</u> _____

_____

____ *Reteaching with Practice* worksheet (specify exercises) _____

____ *Personal Student Tutor* for Lesson 9.7

Real-Life Application: When Will I Ever Use This?

For use with pages 548–553

Fishing

Fishing is one of the most popular, relaxing, and rewarding forms of outdoor recreation. People enjoy fishing in a wide variety of lakes, oceans, rivers, and streams.

Some common methods of fishing include casting, still fishing, drift fishing, trolling, and ice fishing. In casting, anglers use rods to throw a line with natural or artificial bait into the water. In still fishing, the angler casts the bait from a bank or an anchored boat and waits for the fish to bite. In drift fishing, the angler allows the bait to trail the boat, which drifts freely in the current. In trolling, the bait is dragged, at or below the surface of the water, behind a moving boat. In ice fishing, the angler fishes through a hole chopped in the ice.

Manufacturers produce a wide variety of tackle (equipment) designed for every type of fishing. Fishing tackle includes rods, reels, lines, leaders, sinkers, floats, hooks, and bait. The choice of equipment depends on the kinds of fish sought.

In Exercises 1-3, use the following information.

You go fishing on a river. This is your first time fishing from a boat, so you bring plenty of tackle to adjust to the fishing conditions of the river. The cross section showing the bottom of the river at its deepest point can be modeled by the equation $y = .0581x^2 - 2.3238x$, where y is the depth of the river (in feet) and x is the horizontal distance from the bank of the river (in feet).

1. Write a system of quadratic inequalities that describe the cross section of water of the river. Assume that the surface of the river is modeled by $y = 0$. Graph the system.

2. Your boat is directly above the deepest point the river. You drop your line with a sinker at the end. Your line is 25 feet long. Will your sinker touch the bottom of the river? Explain your reasoning.

3. There is a fish detector on the boat. The detector shows that there is a school of fish about 18 feet directly below you. You are 8 feet from the bank of the river. Is the fish detector accurate? Explain your reasoning.

Challenge: Skills and Applications

For use with pages 548–553

1. Sketch the graph of the inequality $y \leq 8 - x^2$.

2. Use the graph from Exercise 1 to determine the solution region for the system of inequalities. Sketch the graph and shade the solution region.

   $y \leq 8 - x^2$

   $y > x + 2$

3. Is the origin in the solution region from Exercise 2?

4. Is the point $(-2, 3)$ in the solution region from Exercise 2?

5. Is the point $(-3, -1)$ in the solution region from Exercise 2?

6. Is the point $(1, 2)$ in the solution region from Exercise 2?

In Exercises 7–11, use the following information.

A company's revenue R (in thousands of dollars) can be modeled by the inequality $R \leq -2x^2 + 60x + 350$, where x is the number of units sold in thousands. The company's costs C can be modeled by the inequality $C \geq 4x + 600$ for x thousand units sold.

7. Draw a graph of the revenue inequality.

8. If the company sells 20 thousand units, can its revenue be $700 thousand?

9. If the company sells 10 thousand units, can its revenue be $750 thousand?

10. Graph the revenue and cost inequalities on the same coordinate grid and shade the region that is a solution to both.

11. Interpret the shaded region for Exercise 10 in terms of the situation.

TEACHER'S NAME _____ CLASS _____ ROOM _____ DATE _____

Lesson Plan

2-day lesson (See *Pacing the Chapter,* TE pages 500C–500D)   **For use with pages 554–560**

GOALS 1. **Choose a model that best fits a collection of data.**
2. **Use models in real-life settings.**

State/Local Objectives _____

_____

✓ Check the items you wish to use for this lesson.

STARTING OPTIONS
_____ Homework Check: TE page 551; Answer Transparencies
_____ Warm-Up or Daily Homework Quiz: TE pages 554 and 553, CRB page 111, or Transparencies

TEACHING OPTIONS
_____ Lesson Opener (Application): CRB page 112 or Transparencies
_____ Graphing Calculator Activity with Keystrokes: CRB pages 113–115
_____ Examples: Day 1: 1–2, SE pages 554–555; Day 2: 2, SE page 556
_____ Extra Examples: Day 1: TE page 555 or Transp.; Day 2: TE page 556 or Transp.
_____ Closure Question: TE page 556
_____ Guided Practice: SE page 557; Day 1: Exs. 1–8; Day 2: none

APPLY/HOMEWORK
Homework Assignment
_____ Basic   Day 1: 9–19; Day 2: 20–22, 27–29, 34–48 even; Quiz 3: 1–13
_____ Average   Day 1: 9–19; Day 2: 20–22; 27–29, 34–48 even; Quiz 3: 1–13
_____ Advanced   Day 1: 9–19; Day 2: 20–22, 27–32, 34–48 even; Quiz 3: 1–13

Reteaching the Lesson
_____ Practice Masters: CRB pages 116–118 (Level A, Level B, Level C)
_____ Reteaching with Practice: CRB pages 119–120 or Practice Workbook with Examples
_____ Personal Student Tutor

Extending the Lesson
_____ Cooperative Learning Activity: CRB page 122
_____ Applications (Interdisciplinary): CRB page 123
_____ Challenge: SE page 559; CRB page 124 or Internet

ASSESSMENT OPTIONS
_____ Checkpoint Exercises: Day 1: TE page 555 or Transp.; Day 2: TE page 556 or Transp.
_____ Daily Homework Quiz (9.8): TE page 560 or Transparencies
_____ Standardized Test Practice: SE page 559; TE page 560; STP Workbook; Transparencies
_____ Quiz (9.7–9.8): SE page 560

Notes _____

_____

_____

TEACHER'S NAME _____ CLASS _____ ROOM _____ DATE _____

Lesson Plan for Block Scheduling

1-day lesson (See *Pacing the Chapter,* TE pages 500C–500D)   **For use with pages 554–560**

GOALS 1. **Choose a model that best fits a collection of data.**
2. **Use models in real-life settings.**

State/Local Objectives _____

_____

_____

✓ Check the items you wish to use for this lesson.

STARTING OPTIONS

____ Homework Check: TE page 551; Answer Transparencies

____ Warm-Up or Daily Homework Quiz: TE pages 554 and
          553, CRB page 111, or Transparencies

TEACHING OPTIONS

____ Lesson Opener (Application): CRB page 112 or Transparencies

____ Graphing Calculator Activity with Keystrokes: CRB pages 113–115

____ Examples: Day 6: 1–2, SE pages 554–555; Day 7: 3, SE page 556

____ Extra Examples: Day 6: TE page 555 or Transp.; Day 7: TE page 556 or Transp.

____ Closure Question: TE page 556

____ Guided Practice: SE page 557; Day 6: Exs. 1–8; Day 7: none

APPLY/HOMEWORK

Homework Assignment (See also the assignment for Lesson 9.7)

____ Block Schedule: Day 6: 9–19; Day 7: 20–22, 27–29, 34–48 even; Quiz 3: 1–13

Reteaching the Lesson

____ Practice Masters: CRB pages 116–118 (Level A, Level B, Level C)

____ Reteaching with Practice: CRB pages 119–120 or Practice Workbook with Examples

____ Personal Student Tutor

Extending the Lesson

____ Cooperative Learning Activity: CRB page 122

____ Applications (Interdisciplinary): CRB page 123

____ Challenge: SE page 559; CRB page 124 or Internet

ASSESSMENT OPTIONS

____ Checkpoint Exercises: Day 6: TE page 555 or Transp.; Day 7: TE page 556 or Transp.

____ Daily Homework Quiz (9.8): TE page 560 or Transparencies

____ Standardized Test Practice: SE page 559; TE page 560; STP Workbook; Transparencies

____ Quiz (9.7–9.8): SE page 560

Notes _____

_____

_____

CHAPTER PACING GUIDE	
Day	**Lesson**
1	Assess Ch. 8; 9.1 (all)
2	9.2 (all); 9.3 (begin)
3	9.3 (end); 9.4 (all)
4	9.5 (all)
5	9.6 (all)
6	9.7 (all); **9.8 (begin)**
7	**9.8 (end)**; Review Ch. 9
8	Assess Ch. 9; 10.1 (all)

NAME _____ DATE _____

WARM-UP EXERCISES

For use before Lesson 9.8, pages 554–560

Find the ratio of the value of the function when $x = 2$ to the value when $x = 1$.

1. $y = 2x^2$

2. $y = 3.5x^2$

3. $y = 3^x$

4. $y = 100(1 - 0.05)^x$

DAILY HOMEWORK QUIZ

For use after Lesson 9.7, pages 548–553

1. Is the ordered pair $(-2, 4)$ a solution of the inequality $y > 3x^2 + 7x + 5$?

2. Which inequality describes the graph?

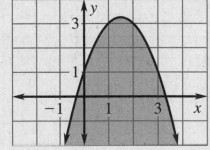

   A. $y > -x^2 + 3x + 1$

   B. $y < -x^2 + 3x + 1$

   C. $y \leq -x^2 + 3x + 1$

3. Graph $y < x^2 - 2x - 2$.

4. The area under a parabolic bridge support is described by $y \leq -0.005x^2 + 120$. The support is 310 ft wide at its base. A road crosses the arch 45 ft in from each side of the base of the support. How high is the road above the base?

NAME _____ DATE _____

Application Lesson Opener

For use with pages 554–560

1. The graph at the right is a scatter plot of the distance Thea traveled in certain periods of time. Do you think this data can be modeled by a *linear* function, an *exponential* function, or a *quadratic* function? Explain.

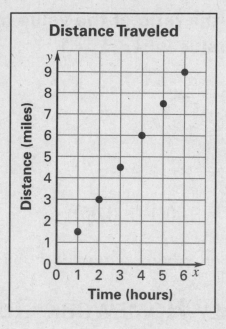

Distance Traveled

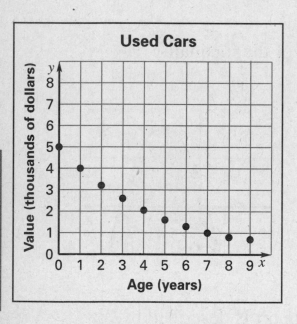

Used Cars

2. The graph at the left is a scatter plot of the value of a used car a certain number of years after the used car was sold. Do you think this data can be modeled by a *linear* function, an *exponential* function, or a *quadratic* function? Explain.

3. The graph at the right is a scatter plot of the height of a ball after a certain amount of time. Do you think this data could be modeled by a *linear* function, an *exponential* function, or a *quadratic* function? Explain.

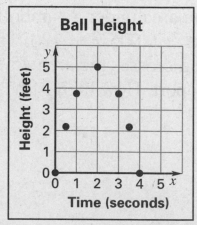

Ball Height

NAME _____ DATE _____

Graphing Calculator Activity

For use with pages 554–560

GOAL To determine whether a linear, exponential, or quadratic model best fits the data

In 1990 you received $50 as a birthday gift. You deposited it into a bank account and made no further deposits or withdrawls. The table below shows the account balance for several years.

Years since 1990, x	0	5	10	15	20	25	30
Balance, y	$50	$62.31	$77.65	$96.76	$120.59	$150.27	$187.27

Activity

❶ Use a graphing calculator to enter the data from the table above and draw a scatter plot. Let L1 represent years since 1990 (x) and L2 represent the account balance (y).

❷ Use linear regression to find a linear model for the data. Round to the nearest thousandth.

❸ Use exponential regression to find an exponential model for the data. Round to the nearest thousandth.

❹ Use quadratic regression to find a quadratic model for the data. Round to the nearest thousandth.

❺ Which model best represents the data? Explain.

Exercises

In Exercises 1–3, decide whether the best model for the data is linear, exponential, or quadratic. Then give the regression equation. Approximate to the nearest thousandth.

1.

°F	−4	14	32	50	68	86
°C	−20	−10	0	10	20	30

2.

Price	$6	$8	$10	$12	$14	$16
Profit	$48	$68	$80	$84	$80	$68

3.

Altitude (thousand ft)	0	5	10	15	20	25	30
Pressure (lb/in.²)	14.7	12.2	10.1	8.3	6.8	5.4	4.4

See page 114 for keystrokes.

Graphing Calculator Activity

For use with pages 554–560

TI-82

STAT 1

Enter years since 1990 in L1.

0 ENTER 5 ENTER 10 ENTER 15

ENTER 20 ENTER 25 ENTER 30 ENTER

Enter account balance in L2.

50 ENTER 62.31 ENTER 77.65 ENTER

96.76 ENTER 120.59 ENTER 150.27 ENTER

187.27 ENTER 2nd [STAT PLOT] 1

Choose the following.

On; Type: ⋰ ;Xlist: L1; Ylist: L2; Mark:▪

WINDOW ENTER 0 ENTER 35 ENTER 5 ENTER

0 ENTER 200 ENTER 20 ENTER GRAPH

STAT ▶ 5 2nd [L1] , 2nd [L2] ENTER

STAT ▶ ALPHA [A] 2nd [L1] , 2nd

[L2] ENTER

STAT ▶ 6 2nd [L1] , 2nd [L2] ENTER

TI-83

STAT 1

Enter years since 1990 in L1.

0 ENTER 5 ENTER 10 ENTER 15

ENTER 20 ENTER 25 ENTER 30 ENTER

Enter account balance in L2.

50 ENTER 62.31 ENTER 77.65 ENTER

96.76 ENTER 120.59 ENTER 150.27 ENTER

187.27 ENTER 2nd [STAT PLOT] 1

Choose the following.

On; Type: ⋰ ;Xlist: L1; Ylist: L2; Mark:▪

WINDOW 0 ENTER 35 ENTER 5

ENTER 0 ENTER 200 ENTER 20

ENTER GRAPH

STAT ▶ 4 2nd [L1] , 2nd [L2] ENTER

STAT ▶ 0 2nd [L1] , 2nd [L2] ENTER

STAT ▶ 5 2nd [L1] , 2nd [L2] ENTER

SHARP EL-9600c

STAT [A] ENTER

Enter years since 1990 in L1.

0 ENTER 5 ENTER 10 ENTER 15

ENTER 20 ENTER 25 ENTER 30 ENTER

Enter account balance in L2.

50 ENTER 62.31 ENTER 77.65 ENTER 96.76

ENTER 120.59 ENTER 150.27 ENTER 187.27

ENTER 2ndF [STAT PLOT] [A] ENTER

Choose the following. on; DATA XY;

ListX: L1; ListY: L2 2ndF [STAT PLOT] [G] 3

WINDOW 0 ENTER 35 ENTER 5 ENTER 0 ENTER

200 ENTER 20 ENTER GRAPH 2ndF [QUIT]

STAT [D] 0 2 (2ndF [L1] , 2ndF [L2]

) ENTER STAT [D] 0 9 (2ndF [L1]

, 2ndF [L2]) ENTER STAT [D] 0 4

(2ndF [L1] , 2ndF [L2]) ENTER

CASIO CFX-9850GA PLUS

From the main menu, choose STAT.

STAT [A] EXE

Enter years since 1990 in List 1.

0 EXE 5 EXE 10 EXE 15 EXE 20 EXE 25 EXE 30

Enter account balance in List 2.

50 EXE 62.31 EXE 77.65 EXE 96.76

EXE 120.59 EXE 150.27 EXE 187.27 EXE F1 F6

Choose the following.

Graph Type: Scatter; XList: List 1; YList: List 2;
Frequency: 1; Mark Type:▪

EXIT SHIFT F3 0 EXE 35 EXE 5 EXE 0 EXE

200 EXE 20 EXE EXIT F1 F1

F1 F6 F3 F6 F6 F2

Do the following to calculate the value of the base
for the exponential model.

MENU 1 SHIFT [e^x] VARS F3 F3 F2

Lesson 9.8

Graphing Calculator Activity Keystrokes

For use with page 559

Keystrokes for Exercise 30

TI-82

STAT 1

Enter x-values in L1.

(-) 3 ENTER (-) 2 ENTER (-) 1 ENTER
0 ENTER 1 ENTER 2 ENTER 3 ENTER

Enter y-values in L2.

4 ENTER 7 ÷ 2 ENTER 3 ENTER
5 ÷ 2 ENTER 2 ENTER 3 ÷ 2 ENTER
1 ENTER

WINDOW ENTER

(-) 5 ENTER 5 ENTER 1 ENTER 0 ENTER
5 ENTER 1 ENTER

2nd [STAT PLOT] 1

On; Type ⋰ ; Xlist: L1; Ylist: L2; Mark: ▫

GRAPH

TI-83

STAT 1

Enter x-values in L1.

(-) 3 ENTER (-) 2 ENTER (-) 1 ENTER
0 ENTER 1 ENTER 2 ENTER 3 ENTER

Enter y-values in L2.

4 ENTER 7 ÷ 2 ENTER 3 ENTER
5 ÷ 2 ENTER 2 ENTER 3 ÷ 2 ENTER
1 ENTER

WINDOW

(-) 5 ENTER 5 ENTER 1 ENTER 0 ENTER
5 ENTER 1 ENTER

2nd [STAT PLOT] 1

On; Type ⋰ ; Xlist: L1; Ylist: L2; Mark: ▫

GRAPH

SHARP EL-9600c

STAT [A] ENTER

Enter x-values in L1.

(-) 3 ENTER (-) 2 ENTER (-) 1 ENTER
0 ENTER 1 ENTER 2 ENTER 3 ENTER

Enter y-values in L2.

4 ENTER 7 ÷ 2 ENTER 3 ENTER
5 ÷ 2 ENTER 2 ENTER 3 ÷ 2 ENTER
1 ENTER

WINDOW

(-) 5 ENTER 5 ENTER 1 ENTER 0 ENTER
5 ENTER 1 ENTER

2ndF [STAT PLOT] [A] ENTER

on; DATA XY; ListX: L1; ListY: L2

2ndF [STAT PLOT] [G] 3

GRAPH

CASIO CFX-9850GA PLUS

From the main menu, choose STAT.

Enter x-values in List 1.

(-) 3 EXE (-) 2 EXE (-) 1 EXE 0 EXE 1
EXE 2 EXE 3 EXE

Enter y-values in L2.

4 EXE 7 ÷ 2 EXE 3 EXE 5 ÷ 2 EXE 2
EXE 3 ÷ 2 EXE 1 EXE

SHIFT F3

(-) 5 EXE 5 EXE 1 EXE 0 EXE
5 EXE 1 EXE

EXIT F1 F6

Graph Type: Scatter; Xlist: List 1; Ylist: List 2;
Frequency: 1; Mark Type: ▫

EXIT F1

Practice A
For use with pages 554–560

Name the type of model described by each equation.

1. $y = 4(1.05)^x$

2. $y = 4x^2 - x + 5$

3. $y = 4x - 5$

4. $y = 3(x - 1)$

5. $y = 3(1.03)^x$

6. $y = 3x^2 - x$

Make a scatter plot of the data. Then name the type of model that best fits the data.

7. $(-1, 1), (0, 3), (1, 5), (2, 7), (3, 9), (4, 11)$

8. $(-3, 2), (-2, -1), (-1, -2), (0, -1), (1, 2), (2, 7)$

9. $(-3, -1), (-2, 0), (-1, 1), (0, 2), (1, 3), (2, 4)$

10.

x	y
-3	$\frac{1}{8}$
-2	$\frac{1}{4}$
-1	$\frac{1}{2}$
0	1
1	2
2	4

11.

x	y
0	2
1	6
2	8
3	8
4	6
5	2

12.

x	y
-2	12
-1	6
0	3
1	$\frac{3}{2}$
2	$\frac{3}{4}$
3	$\frac{3}{8}$

13. *Multiple Choice*   The graph shows the percentage of recording shipments that are music cassettes from 1990 to 1997. Which type of model best fits the data?

A. Exponential   **B.** Quadratic

C. Linear   **D.** None

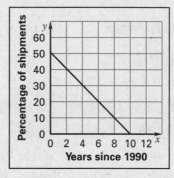

14. *Writing*   From Exercise 13, why do you think the number of shipments of cassettes is decreasing the way it is?

15. *Simple Interest*   The balance B in an account that pays 5% interest for t years is shown in the table. Which type of model best fits the data? Write a model.

Time, t	0	1	2	3	4	5
Balance, B	100	105	110	115	120	125

16. *Surface Area of a Cube*   The surface area A of a cube with side s is shown in the table. Find a model that relates the surface area and side.

Side, s	1	2	3	4	5	6
Surface Area, A	6	24	54	96	150	216

Lesson 9.8

NAME _____ DATE _____

Practice B

For use with pages 554–560

Write an equation for each type of algebraic model named. Sketch a graph of your model.

  1. linear   2. exponential   3. quadratic

Make a scatter plot of the data. Then name the type of model that best fits the data.

  4. $(-1, -16), (2, -0.25), (1, -1), (0, -4), (2.5, -0.125), (0.5, -2)$

  5. $\left(1, \frac{3}{2}\right), \left(\frac{1}{2}, \frac{1}{8}\right), \left(-1, -\frac{5}{2}\right), (2, 5), \left(-3, -\frac{5}{2}\right), (-2, -3)$

  6. $(2, -16), (-1, 2), (3, -22), \left(\frac{1}{2}, -7\right), (0, -4), \left(-\frac{3}{4}, \frac{1}{2}\right)$

7.

x	y
-4	-14
$-\frac{5}{2}$	-8
-1	-2
$\frac{1}{2}$	4
2	10
3.5	16

8.

x	y
-3	-22
-2	-11
-1	-4
0	-1
1	-2
2	-7

9.

x	y
-4	32
0	2
-2	8
5	$\frac{1}{16}$
2	$\frac{1}{2}$
-1	4

10. **Kinetic Energy**   The kinetic (moving) energy E (in joules) of a 10 kg object moving with velocity v (in meters per second) is shown in the table. Which type of model best fits the data? Write the model.

Velocity, v	0	5	10	15	20	25
Kinetic Energy, E	0	125	500	1125	2000	3125

Cell Sites   **In Exercises 11 and 12, use the following information.**

The number of cell sites used in the cellular phone industry in the United States from 1990 to 1995 can be modeled by

$$C = 5784.22(1.318)^t$$

where $t = 0$ represents 1990.

11. Copy and complete the table.

Year t	Cell Sites C	Change in C from previous year
0	5784.22	—
1		
2		
3		
4		
5		

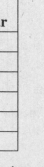

12. Describe the pattern in the change of cell sites.

Practice C

For use with pages 554–560

Make a scatter plot of the data. Then name the type of model that best fits the data.

1. $(1.5, 4.25), (4, 8), (-3, -2.5), (-1, 0.5), (0, 2), (3, 6.5)$

2. $(0, 3), (-2, 12), \left(2, \frac{3}{4}\right), \left(1, \frac{3}{2}\right), \left(3, \frac{3}{8}\right), (-1, 6)$

3. $\left(-2, \frac{8}{9}\right), \left(-1, \frac{4}{3}\right), (0, 2), (1, 3), \left(2, \frac{9}{2}\right), \left(3, 6\frac{3}{4}\right)$

4.

x	y
3	13
1	-3
4	27
-3	13
-1	-3
0	-5

5.

x	y
2	-3
$\frac{4}{5}$	$\frac{27}{25}$
-1	$4\frac{1}{2}$
0	3
-3	4.5
-5	0.5

6.

x	y
1	-3.2
0	-4
-1	-5
-2	-6.25
2	-2.56
-3	$-7\frac{13}{16}$

7. *Depreciation* The value V of a piece of equipment between 1990 and 1995 is given in the table. Let t represent the year with $t = 0$ corresponding to 1990. Find a model that relates the year and value.

Time, t	0	1	2	3	4	5
Value, V ($)	500	450	400	350	300	250

Population **In Exercises 8 and 9, use the following information.**

The resident population P (in millions) of the United States since 1900 can be modeled by

$$P = 0.0090t^2 + 1.1121t + 77.9958$$

where t is the number of years after 1990.

8. Copy and complete the table.

Year t	Population P	Year t	Population P
0	77,995,800	50	
10		60	
20		70	
30		80	
40		90	

9. Describe the pattern in the change of population.

10. *Extension* Make a scatter plot of the data. Name the type of model that best fits the data. $(-3, 3), (-2, 2), (-1, 1), (0, 0), (1, 1), (2, 2), (3, 3)$

NAME _____ DATE _____

Reteaching with Practice

For use with pages 554–560

GOAL Choose a model that best fits a collection of data and use models in real-life settings

---

VOCABULARY

Linear Model	Exponential Model	Quadratic Model
$y = mx + b$	$y = C(1 \pm r)^t$	$y = ax^2 + bx + c$

EXAMPLE 1 *Choosing a Model*

Name the type of model that best fits each data collection.

a. $\left(-2, \frac{1}{4}\right), \left(-1, \frac{1}{2}\right), (0, 1), (1, 2), (2, 4)$

b. $(-2, -3), (-1, -1), (0, 1), (1, 3), (2, 5)$

c. $(-2, 5), (-1, 2), (0, 1), (1, 2), (2, 5)$

SOLUTION

Make scatter plots of the data. Then decide whether the points lie on a line, an exponential curve, or a parabola.

a. Exponential Model   **b.** Linear Model   **c.** Quadratic Model

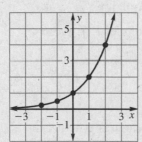   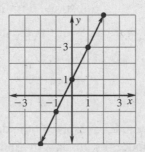   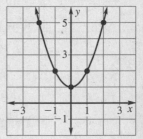

Exercises for Example 1

Make a scatter plot of the data. Then name the type of model that best fits the data.

1. $(-2, 3), (-1, 0), (0, -1), (1, 0), (2, 3)$

2. $(-2, -3), (-1, -2), (0, -1), (1, 0), (2, 1)$

3. $(-2, 4), (-1, 2), (0, 1), \left(1, \frac{1}{2}\right), \left(2, \frac{1}{4}\right)$

Reteaching with Practice

For use with pages 554–560

EXAMPLE 2 *Writing a Model*

Your biology class is studying the population growth of fruit flies. The table shows the population, P (number of fruit flies) for various times, t (in weeks). Which type of model best fits the data?

t	0	1	2	3	4
P	2	6	18	54	162

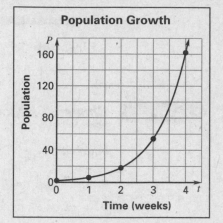

Population Growth

SOLUTION

Draw a scatter plot of the data. You can see that the graph is curved, not linear. Test whether a quadratic model fits. Begin by writing the simple quadratic model $P = at^2$. To find a, substitute any known values of P and t.

$P = at^2$	Write quadratic model.
$6 = a \cdot 1^2$	Substitute 6 for P and 1 for t.
$6 = a$	Solve for a.
$P = 6t^2$	Substitute 6 for a in the equation.

Now check several values of t in the quadratic model $P = 6t^2$.

$P = 6t^2$	$P = 6t^2$
$P = 6(2)^2$	$P = 6(4)^2$
$P = 24 \neq 18$	$P = 96 \neq 162$

The quadratic model does not fit the data. You can test whether an exponential model fits the data by finding the ratios of consecutive populations.

$$\frac{\text{Population Week 0}}{\text{Population Week 1}} = \frac{2}{6} = \frac{1}{3}$$

$$\frac{\text{Population Week 1}}{\text{Population Week 2}} = \frac{6}{18} = \frac{1}{3}$$

Because the populations increase by the same percent, an exponential model fits the data.

Exercise for Example 2

4. Which type of model best fits the data?

t	0	1	2	3	4
P	2	3	6	11	18

NAME _____ DATE _____

Quick Catch-Up for Absent Students

For use with pages 554–560

The items checked below were covered in class on (date missed) _____

Lesson 9.8: Comparing Linear, Exponential, and Quadratic Models

____ **Goal 1:** Choose a model that best fits a collection of data. (p. 554)

Material Covered:

____ Example 1: Choose a Model

____ **Goal 2:** Use models in real-life settings. (pp. 555–556)

Material Covered:

____ Student Help: Look Back

____ Example 2: Writing a Model

____ Example 3: Writing a Model

____ Other (specify) _____

_____

Homework and Additional Learning Support

____ Textbook (specify) <u>pp. 557–560</u>_____

_____

____ *Reteaching with Practice* worksheet (specify exercises)_____

____ *Personal Student Tutor* for Lesson 9.8

Cooperative Learning Activity

For use with pages 554–560

GOAL **To research the lives of some prominent mathematicians**

Materials: research materials

Exploring Famous Mathematicians

In this chapter, you have been introduced to several important algebraic concepts. In this activity, you will research one of the mathematicians who developed these concepts.

Instructions

① Choose one of the mathematicians from the list below.

② Conduct research on this individual.

③ Write a paper discussing the life and the contributions of the individual and present your findings to the class.

Niccolo Tartaglia	Isaac Newton	Rene Descartes
Omar Khayyam	Maria Agnesi	Leonard of Pisa (Fibonacci)
Mary Somerville	Neils Abel	Albert Einstein
Hypatia	Evariste Galois	Emmy Noether

Analyzing the Results

1. What were the most significant contributions of these mathematicians? What were some of their greatest obstacles?

2. Do you know of any modern-day mathematicians? What are they currently researching?

NAME _____ DATE _____

Interdisciplinary Application

For use with pages 554–560

Ants

BIOLOGY   Ant is the name of a family of small insects that live in organized communities. Ants are also known as social insects. A community of social insects is called a colony. An ant colony may have dozens, hundreds, thousands, or millions of members.

In biology class you set up three ant colonies. Each colony starts with one queen ant and one male ant. Every other Friday (every 14 days) you estimate the number of ants in each colony. The tables below represent the three colonies.

Colony A	
Day, x	Ants, y
0	2
14	287
28	434
42	445
56	318
70	54

Colony B	
Day, x	Ants, y
0	2
14	82
28	162
42	241
56	321
70	401

Colony C	
Day, x	Ants, y
0	2
14	9
28	37
42	160
56	690
70	2976

1. Make a scatter plot for each colony.

2. Choose the type of model (linear, exponential, or quadratic) that best fits the data for each colony.

3. Write an equation that best fits the scatter plot for each colony.

NAME _____ DATE _____

Challenge: Skills and Applications

For use with pages 554–560

In Exercises 1–6, use the following data.

$(0, 1), (2, 2), (3, 4), (5, k)$

1. Find a value of k that makes the data approximately fit a linear model.

2. Write a linear model for the data with the value of k from Exercise 1.

3. Find a value of k that makes the data approximately fit a quadratic model.

4. Write a quadratic model for the data with the value of k from Exercise 3.

5. Find a value of k that makes the data approximately fit an exponential model.

6. Write an exponential model for the data with the value of k from Exercise 5.

In Exercises 7–11, use the table which shows the exchange rates between the United States dollar and the Indian rupee from 1970 to 1995. Let x be the number of 5-year intervals after 1970.

Year	1970	1975	1980	1985	1990	1995
Rupees per dollar	7.576	8.409	7.887	12.369	17.504	32.427

7. Use the 1970 and 1975 exchange rates to write an exponential model of the data.

8. Evaluate the model from Exercise 7 for $x = 5$. How well does it fit the data?

9. Use the 1970 and 1975 exchange rates to write a quadratic model of the data in the table.

10. Evaluate the model from Exercise 9 for $x = 5$. How well does it fit the data?

11. According to the results from Exercises 8 and 10, which model seems to fit the data better?

NAME _____ DATE _____

Chapter Review Games and Activities

For use after Chapter 9

Crossword puzzle: Using the clues at the bottom of the page, review vocabulary from Chapter 9.

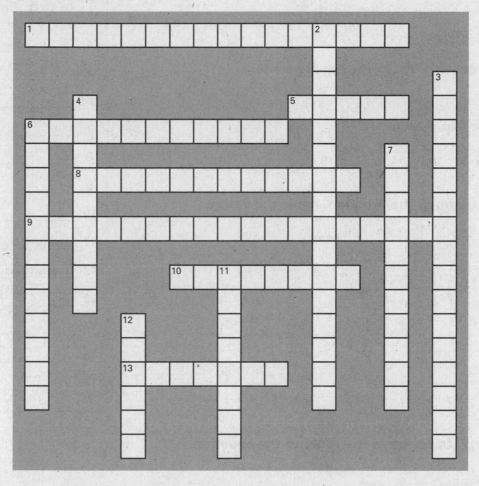

Across

1. $\sqrt{\dfrac{a}{b}} = \dfrac{\sqrt{a}}{\sqrt{b}}$

5. The solutions of a quadratic equation

6. All positive real numbers have two of these

8. $b^2 - 4ac$ in the quadratic formula

9. a in $ax^2 + bx + c = 0$

10. The graph of a quadratic function

13. $\sqrt{}$ (the square root symbol)

Down

2.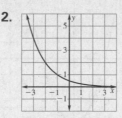

3. $x = \dfrac{-b \pm \sqrt{b^2 - 4ac}}{2a}$

4. An equation whose standard form is $ax^2 + bx + c = 0$

6. No perfect square factors in the radicand
   No fractions in the radicand
   No radicals appear in the denominator of a fraction

7.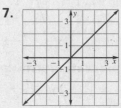

11. 27 in $\sqrt{27}$

12. The lowest or highest point on a parabola

Algebra 1
Chapter 9 Resource Book

125

Review and Assess

NAME _____ DATE _____

Chapter Test A

For use after Chapter 9

Evaluate the expression.

1. $\sqrt{81}$ 2. $-\sqrt{100}$

3. $\sqrt{b^2 - 4ac}$ when $a = 3, b = 7, c = 2$

4. $\sqrt{b^2 - 4ac}$ when $a = 3, b = 8, c = 4$

Solve the equation by finding square roots.

5. $x^2 = 81$ 6. $x^2 = 49$

Simplify the expression.

7. $\sqrt{45}$ 8. $\sqrt{54}$ 9. $\sqrt{\dfrac{16}{25}}$ 10. $\sqrt{\dfrac{10}{32}}$

Sketch the graph of the function. Label the vertex.

11. $y = 3x^2$ 12. $y = -x^2$

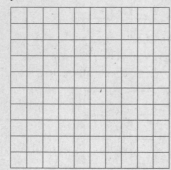 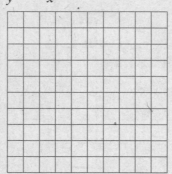

| 1. _____ |
| 2. _____ |
| 3. _____ |
| 4. _____ |
| 5. _____ |
| 6. _____ |
| 7. _____ |
| 8. _____ |
| 9. _____ |
| 10. _____ |
| 11. Use grid at left. |
| 12. Use grid at left. |
| 13. _____ |
| 14. _____ |
| 15. _____ |
| 16. _____ |

Use the graph to estimate the roots of the equation.

13. $y = x^2 - 2x - 8$ 14. $y = x^2 - 4$

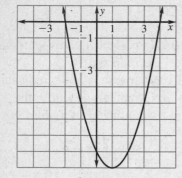 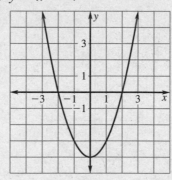

Use the quadratic formula to solve the equation.

15. $0 = x^2 + x - 20$ 16. $0 = x^2 - 5x + 6$

Find the x-intercepts of the graph of the equation.

17. $y = x^2 - 3x + 2$ 18. $y = x^2 - 1$

Decide how many solutions the equation has.

19. $x^2 - 2x + 1 = 0$ 20. $x^2 + 3 = 0$

Sketch the graph of the inequality.

21. $y \geq x^2$ 22. $y < x^2 - 3$

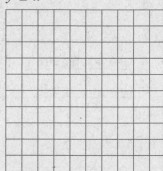   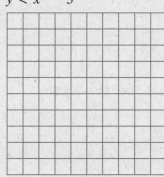

17. _____

18. _____

18. _____

19. _____

20. _____

21. _____

22. _____

23. _____

24. _____

25. _____

26. _____

23. The revenue from selling x units of a product is given by
$y = -0.0002x^2 + 20x$. How many units must be sold in order to
have the greatest revenue? (Find the x-coordinate of the vertex of
the parabola.)

Name the type of model that best fits the data.

24.
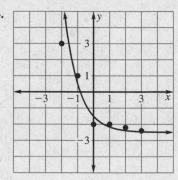

25.
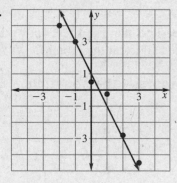

26.
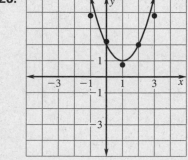

Algebra 1
Chapter 9 Resource Book

127

Review and Assess

Chapter Test B

For use after Chapter 9

Evaluate the expression.

1. $\sqrt{169}$

2. $-\sqrt{625}$

3. $\sqrt{b^2 - 4ac}$ when $a = 4$, $b = 5$, $c = 1$

Solve the equation by finding square roots.

4. $x^2 + 4 = 20$

5. $x^2 - 7 = 29$

Simplify the expression.

6. $\sqrt{\dfrac{9}{16}}$

7. $\sqrt{\dfrac{12}{75}}$

Sketch the graph of the function. Label the vertex.

8. $y = x^2 + 2x + 1$

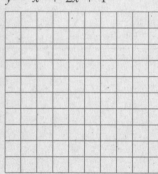

9. $y = x^2 + 4x + 1$

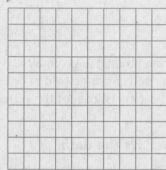

10. You drop a ball from a set of bleachers that is 64 feet above the ground. How long will it take the ball to hit the ground?

Use the graph to estimate the roots of the equation.

11. $y = x^2 - x - 6$

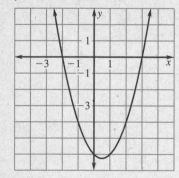

12. $y = x^2 - 6x + 8$

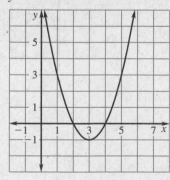

Use the quadratic formula to solve the equation.

13. $0 = x^2 - 2x - 15$

14. $0 = x^2 + 10x + 24$

1.	_____
2.	_____
3.	_____
4.	_____
5.	_____
6.	_____
7.	_____
8.	Use grid at left.
9.	Use grid at left.
10.	_____
11.	_____
12.	_____
13.	_____
14.	_____

Review and Assess

Chapter Test B

For use after Chapter 9

Find the *x*-intercepts of the graph of the equation.

15. $y = x^2 - 12x + 35$

16. $y = x^2 - 7x + 6$

Decide how many solutions the equation has.

17. $x^2 + 4x + 4 = 0$

18. $x^2 - 6x + 13 = 0$

Sketch the graph of the inequality.

19. $y > x^2 - 5x + 4$

20. $y \leq x^2 - 2x + 3$

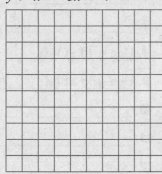

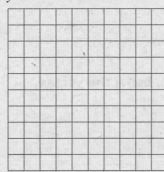

15. _____	
16. _____	
17. _____	
18. _____	
19. _____	
20. _____	
21. _____	
22. _____	
23. _____	
24. _____	

21. The revenue from selling *x* units of a product is given by
$y = -0.0002x^2 + 40x$. How many units must be sold in order to
have the greatest revenue? (Find the *x*-coordinate of the vertex of
the parabola.)

Name the type of model that best fits the data.

22.

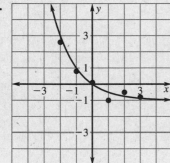

23.

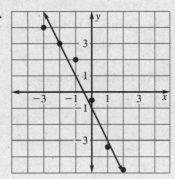

24.

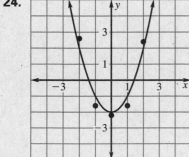

Review and Assess

Chapter Test C

For use after Chapter 9

Evaluate the expression.

1. $\sqrt{0.0625}$ 2. $\pm\sqrt{1.44}$

3. $\sqrt{b^2 - 4ac}$ when $a = 2, b = 6, c = -3$

Solve the equation by finding square roots.

4. $3x^2 - 6 = 21$ 5. $6x^2 - 8 = 46$

Simplify the expression.

6. $\sqrt{\dfrac{12}{25}}$ 7. $\dfrac{\sqrt{12} \cdot \sqrt{16}}{\sqrt{75}}$

Sketch the graph of the function. Label the vertex.

8. $y = -2x^2 + 4x - 1$ 9. $y = \frac{1}{2}x^2 + 2x + 3$

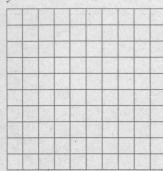   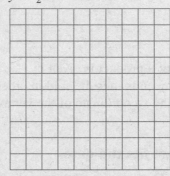

10. You drop a ball from a set of bleachers that is 144 feet above the ground. How long will it take the ball to hit the ground?

Use the graph to estimate the roots of the equation.

11. $y = x^2 + 3x - 10$ 12. $y = x^2 - 6x + 9$

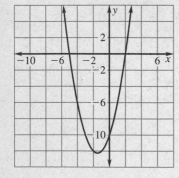   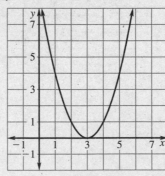

Use the quadratic formula to solve the equation.

13. $0 = x^2 - 8x + 15$ 14. $0 = x^2 + 2x - 24$

1.	_____
2.	_____
3.	_____
4.	_____
5.	_____
6.	_____
7.	_____
8.	**Use grid at left.**
9.	**Use grid at left.**
10.	_____
11.	_____
12.	_____
13.	_____
14.	_____

Chapter Test C

For use after Chapter 9

Find the *x*-intercepts of the graph of the equation.

15. $y = x^2 + 2x - 35$

16. $y = x^2 + 2x - 48$

Decide how many solutions the equation has.

17. $x^2 - 10x + 25 = 0$

18. $x^2 + 8x + 19 = 0$

Sketch the graph of the inequality.

19. $y \geq x^2 + 2x + 3$

20. $y < x^2 - 6x + 5$

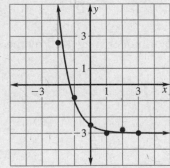

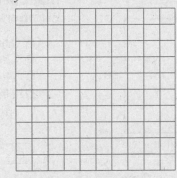

15. _____

16. _____

17. _____

18. _____

19. _____

20. _____

21. _____

22. _____

23. _____

24. _____

21. The revenue from selling *x* units of a product is given by $y = -0.0002x^2 + 60x$. How many units must be sold in order to have the greatest revenue? (Find the *x*-coordinate of the vertex of the parabola.)

Name the type of model that best fits the data.

22.

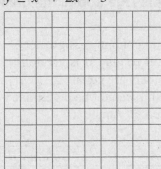

23.

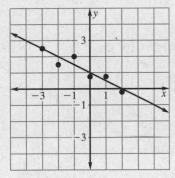

24.

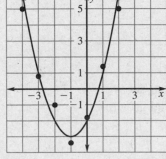

SAT/ACT Chapter Test

For use after Chapter 9

1. Which of the following is a solution of the equation $\frac{3}{4}x^2 - 13 = 14$?

   A $\frac{9}{2}$ **B** $\sqrt{27}$

   C -6 **D** $\frac{2}{\sqrt{3}}$

2. Find the area of the rectangle.

   A 294
   B $7\sqrt{6}$
   C $\sqrt{35}$
   D $7\sqrt{5}$

   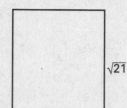

   $\sqrt{21}$

   $\sqrt{14}$

3. Which of the following is the simplified form of $3\dfrac{\sqrt{35}\ \sqrt{15}}{\sqrt{63}}$?

   A $5\sqrt{3}$ **B** $\dfrac{5\sqrt{2}}{\sqrt{7}}$

   C $\dfrac{5\sqrt{3}}{3}$ **D** $3\sqrt{3}$

4. What is the x-coordinate of the vertex for the graph of the equation $y = \frac{2}{3}x^2 - 6x + 4$?

   A $\frac{1}{18}$ **B** $\frac{1}{3}$

   C 3   **D** $\frac{9}{2}$

5. What are the solutions of the equation $y = -x^2 + x + 6$?

   A -2 and 3   **B** $\frac{5}{2}$ and $-\frac{5}{2}$

   C 3 and -5 **D** None

6. What is the discriminant of the equation $y = 3x^2 - 5x + 1$?

   A 13   **B** 29
   C 41   **D** 61

In Questions 7 and 8, choose the statement below that is true about the given numbers.

   A The number in column A is greater.
   B The number in column B is greater.
   C The two numbers are equal.
   D The relationship cannot be determined from the given information.

7.
Column A	Column B
$-\sqrt{0.16}$	$-\sqrt{0.25}$

   A   **B**   **C**   **D**

8.
Column A	Column B
x in $x^2 + 4 = 13$	x in $x^2 + 3 = 19$

   A   **B**   **C**   **D**

9. Name the type of model suggested by the graph.

   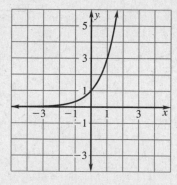

   A Quadratic
   B Exponential decay
   C Exponential growth
   D Linear

JOURNAL **1.** In this chapter, you have learned how to solve vertical motion problems using two models. (a) Create three different word problems: one in which an object is dropped, one in which an object is thrown down, and one where an object is thrown upward. At least two of the three problems should request the solver to find the time traveled in the air. Be sure the objects are thrown in a vertical motion. (b) Solve each problem, showing all work. As part of your solution, explain how you know what number to substitute in for each variable.

MULTI-STEP PROBLEM **2.** During testing, it was determined that the horsepower for an engine is represented by $H = -59.3r^2 + 1041r - 3676$, where H represents the horsepower and r represents revolutions per minute (rev/min) in thousands. (Horsepower is a unit for measuring the power of an engine and other devices.)

For problems a–d, round answers to the nearest whole number.

a. Predict what the graph will look like. Include what shape the graph will be and explain how you know. What does the leading coefficient tell you about the graph?

b. Graph the engine's horsepower equation from 7200 rev/min to 9600 rev/min. (Hint: rev/min is in thousands). Sketch the curve and label your axes. Use the curve to explain how revolutions per minute affects horsepower.

c. Find the horsepower at 8000 rev/min. Explain how you determined your answer.

d. Estimate the rev/min at which the engine produces 876 horsepower. At how many points does this occur? Describe, in two or three complete sentences, how you could solve this problem both by using your graphing calculator and algebraically.

3. *Writing* After some adjustments to the engine, another test is run. The engine now develops a maximum of 902 horsepower. What is the percentage improvement in horsepower? Explain how you got your answer.

Review and Assess

Alternative Assessment Rubric

For use after Chapter 9

JOURNAL SOLUTION

1. **a.** Answers may vary; *Sample answer:* A penny is dropped from the top of a 400 story building. How long will it take to hit the ground? A penny is thrown down at 10 feet per second. This time, how soon will it hit the ground? A model rocket is shot vertically into the air from a height of 10 feet? How high is it after 2 seconds? **b.** Complete answers should address these points:

- The model $h = -16t^2 + s$ is used for a dropped object and $h = -16t^2 + vt + s$ for objects thrown down or up.

- When finding how long it takes an object to hit the ground, 0 is substituted in for h, since the height at ground level is zero.

- Heights chosen other than the ground should be set equal to zero before solving.

- Initial height is substituted in for s, and the initial velocity for v.

- Initial velocity is positive for an object thrown upward and negative for an object thrown downward.

MULTI-STEP PROBLEM SOLUTION

2. **a.** *Sample answer:* the graph would be the shape of a parabola, because the model is quadratic. It would open downward because the leading coefficient is negative.

b. Check graphs. *Sample answer:* As revolutions per minute increase, the horsepower will increase until it reaches its highest point, after which horsepower decreases.

c. 857; substitute 8 in for r in the equation for horsepower and simplify.

d. Estimates should be close to 8248 rev/min and 9307 rev/min. *Sample answer:* to solve on a graphing calculator, use the calculator's trace feature to see when H is close to 876; to solve algebraically substitute 876 in for H, set equation equal to zero, and solve using the quadratic formula.

3. 1%; *Sample answer:* Substitute the r-coordinate into the equation to find the first maximum horsepower to be 893; $\frac{902}{893} \approx 1.01$, a 1% improvement.

MULTI-STEP PROBLEM RUBRIC

4 Students complete all parts of the questions accurately. Students correctly explain how to answer questions both graphically and algebraically. Graph is sketched correctly. They understand that the vertex is the maximum point and correctly calculate the percentage improvement.

3 Students complete the questions and explanations. Solutions may contain minor mathematical errors or misunderstandings. Students may fail to give answers for rev/min in thousands. Graph is sketched accurately. Students understand what the vertex is, and correctly calculate percentage improvement.

2 Students complete questions and explanations. Some mathematical errors may occur. Students fail to correctly explain how to answer questions graphically and algebraically. Graph is incomplete or incorrect. Students have trouble interpreting the graph or fail to explain that the maximum point is the vertex.

1 Answers are incomplete. Solutions and explanations are incorrect. Graph is missing or completely inaccurate.

Algebra 1
Chapter 9 Resource Book

CHAPTER

9

Project: Light Square

For use with Chapter 9

OBJECTIVE Determine the relationship between the area covered by a light from a limited source to the distance of the area from the source.

MATERIALS flashlight, thick cardboard, electrical tape, ruler, scissors, graph paper

INVESTIGATION Cut a square hole with a side length of $\frac{1}{2}$ inch near the center of the piece of cardboard. Reinforce the edges of the hole with the electrical tape. Tape a sheet of graph paper on a wall in a dark room. Hold the cardboard 6 inches from the graph paper and parallel to it. Hold the flashlight close to the cardboard and shine the flashlight through the hole. Make sure the flashlight shines only through the hole in the cardboard and not around the edges of it. Mark the corners of the square of light formed on the graph paper.

1. Find the area of the square of light formed when the cardboard is held 6 inches from the graph paper.

2. Repeat the activity to find the area of the square of light formed when the cardboard is held 12 inches from the graph paper.

3. Write an equation of the form $y = kd^2$, where y is the area of the square when the cardboard is d inches from the light. Let k be the average of the two constants of variation you found with each data point.

4. Use your equation to predict the area of the light formed when the cardboard is held 9 inches from the graph paper. Test your prediction.

5. Formulate a conjecture about the area covered by a light in relation to the distance from a limited source.

PRESENT YOUR RESULTS Make a poster presenting the results of your experiment. Include diagrams, an explanation of the procedures you used, your data, your equations, and your conjecture.

Review and Assess

Project: Teacher's Notes

For use with Chapter 9

GOALS
- Use a ratio to model a situation involving the square of data.
- Use quadratic models in real-life settings.
- Use mathematical models to make predictions.

MANAGING THE PROJECT
You may wish to have students work in small groups, especially to collect the data. If the square of light is not distinct or clearly square, students may need to interpolate to determine the dimensions of the square. Important issues to address are: the meaning of direct variation and how an equation of the form $y = kd^2$ is similar to direct variation. In this case y varies directly as the square of d.

RUBRIC
The following rubric can be used to assess student work.

4 The student collects data carefully and accurately, derives an equation correctly, and makes an accurate prediction. The conjecture is well written and correct. The poster clearly presents the procedures and results of the experiment.

3 The student collects data, derives an equation, and makes a prediction. However, the student may not perform all calculations accurately or may make errors in collecting the data. The poster presents the procedures and results of the experiment, but the presentation may not be as thorough as possible.

2 The student collects data, derives an equation, and makes a prediction. However, work may be incomplete or reflect misunderstanding. For example, the student may not collect the data carefully or accurately or may find a direct linear variation model rather than a quadratic one. The poster may indicate a limited grasp of certain ideas or may lack key explanations.

1 Data, the equation, or the prediction is missing or student does not show an understanding of key ideas. The poster does not present an appropriate procedure or gives inaccurate results.

Algebra 1
Chapter 9 Resource Book

Simplify the expression. (2.7)

1. $45y \div \dfrac{5y}{6}$

2. $-\dfrac{8t}{9} \div \dfrac{9}{8t}$

3. $-65x^2 \div \dfrac{13x}{2}$

4. $-6 \cdot \left(\dfrac{2w}{-6}\right)$

5. $7 \cdot \left(-\dfrac{y}{14}\right)$

6. $\dfrac{18t}{24s} \div \dfrac{9t}{12s}$

Solve. Round to the nearest hundredth. (3.6)

7. $-56y + 57 = 290$

8. $57.1x - 9.2 = -122.3$

9. $2.009s = 3.990 + 3.992s$

10. $4 + 9.6 - 3.89 = 4.12a$

Find the slope and *y*-intercept of the graph of the equation. (4.6)

11. $y = \frac{2}{15}x - \frac{8}{5}$

12. $2x + 6y = 9$

13. $12 - 54x = 60y$

14. $35 = \frac{11}{2}y - x$

15. $4\frac{1}{5}y = 3\frac{1}{5}x$

16. $9.23y = 45$

Write the equation of the line that is perpendicular to the given line and passes through the given point. (5.3)

17. $y = 2x - 8, (0, 2)$

18. $5x - 9y = 10, (-3, 7)$

19. $7y = 5x - 12, (4, -6)$

20. $8x = \frac{8}{11}y - 22, (1, -1)$

Write an equation in standard form of the line that passes through the given point and has the given slope. (5.6)

21. $(3, 5), m = 8$

22. $(4.2, 8), m = -1$

23. $\left(\frac{3}{8}, -1\right), m = -9$

Graph the solution of the inequality. (6.2)

24. $-9 > n - 12$

25. $x + 34 \geq 36$

26. $\dfrac{a}{14} - \dfrac{1}{28} > \dfrac{1}{56}$

27. $3 \leq 6.5 + y$

28. $\dfrac{4}{3c} \geq \dfrac{16}{9}$

29. $12.0 > 13.9 - x$

Choose a method to solve the linear system. (7.1–7.3)

30. $8x + 2y = 16$
$\ 5x - y = 28$

31. $4x - 6y = -6$
$\ 10x + 7y = -4$

32. $3a + 3b = 7$
$\ 3a + 5b = 3$

Simplify, if possible. Write your answer as a power. (8.1)

33. $7^3 - 6^3 - 7^2$

34. $-(4x)^2 \cdot (3x^3)^5$

35. $(-2xy)^4(-xy)^2$

36. $-(r^5s^9)(r^4s^5)^2$

37. $(a^4b^6c^7)(ab)^5$

38. $(-2xy^5)(6y^6)^3$

Evaluate the expression. Give the exact value if possible. Otherwise round to the nearest hundredth. (9.1)

39. $-\sqrt{121}$

40. $\sqrt{15}$

41. $\sqrt{289}$

42. $\sqrt{0.02}$

43. $\pm\sqrt{0.001}$

44. $\pm\sqrt{20.25}$

Review and Assess

Simplify the expression. (9.2)

45. $\sqrt{\frac{3}{16}}$

46. $\sqrt{\frac{8}{25}}$

47. $11\sqrt{\frac{36}{4}}$

48. $\dfrac{\sqrt{46}}{\sqrt{81}}$

49. $\frac{1}{6}\sqrt{48}\cdot\sqrt{3}$

50. $-\dfrac{\sqrt{2}}{\sqrt{5}}\cdot\dfrac{\sqrt{8}}{\sqrt{9}}$

Tell whether the graph opens up or down and find an equation of the axis of symmetry. (9.3)

51. $y = 5x^2 - 4x$

52. $y + 3 = \frac{1}{9}x^2 - 7x$

53. $y = -14x^2 - \frac{5}{9}x + 2$

Solve the equation algebraically. Check the solution graphically. (9.4)

54. $5x^2 = 30$

55. $\frac{1}{3}x^2 = 48$

56. $4x^2 - 10 = 390$

Use the quadratic formula to solve the equation. (9.5)

57. $x^2 - 2x - 3 = 0$

58. $-\frac{1}{2}x^2 - x + 2 = 0$

59. $5x^2 + 2x - 1 = 0$

60. $3x^2 - 7x + 2 = 0$

61. $-c^2 = 3c - 3$

62. $5z = 12 - 4z^2$

Tell whether the equation has *two solutions*, *one solution*, or *no real solution*. (9.6)

63. $3x^2 - 8x + 4 = 0$

64. $-6y^2 + y - 8 = 0$

65. $x^2 - 12x + 36 = 0$

Sketch the graph of the inequality. (9.7)

66. $y < 2x^2 - x$

67. $y \geq x^2 + x - 5$

68. $y < 2x^2 + 3x - 6$

Name the type of model (linear, exponential, or quadratic) that best fits each data collection. (9.8)

69. $(0, 0.5), (1, 0.6), (3, 1.0), (5, 1.8)$

70. $(-2, 5), (0, 1), (1, 2), (2, 5)$

71. $(-1, 2), (0, 0), (1, 2), (2, 8)$

72. $(-2, 2), (0, 4), (2, 6), (4, 8)$

Review and Assess

ANSWERS

Chapter Support

Parent Guide
Chapter 9

9.1: 2, −2 **9.2:** 30 **9.3:** 0.25 sec; 5 ft

9.4: about 0.81 sec **9.5:** $\frac{5}{2}$, −1 **9.6:** −23; none; no **9.7:** not a solution **9.8:** 252 in.^{2}

Prerequisite Skills Review

1. 335 **2.** 270 **3.** $-\frac{20}{3}$ **4.** 2

5.

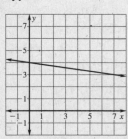

6.

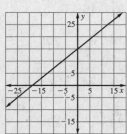

7.

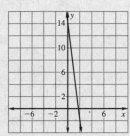

8.

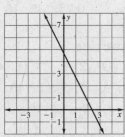

9. yes **10.** no **11.** no **12.** yes

Strategies for Reading Mathematics

1. a. 2704 **b.** 3600 **c.** 2809 **d.** 3249

2. a. 1.732 **b.** 2.449 **c.** 7.141 **d.** 7.616

3. a. 10 **b.** 3 **c.** 8 **d.** 59

Lesson 9.1

Warm-up Exercises

1. 36 **2.** 196 **3.** −81 **4.** 0 **5.** −36

Daily Homework Quiz

1. about $18,820

2. exponential decay; 0.89; 11%

3. exponential growth; 1.03; 3%

4. about 422 mg

Lesson Opener
Allow 10 minutes.

1. 64 ft^2 **2.** 144 ft^2 **3.** *Sample answer:* Find the number that can be multiplied by itself to get the area. **4.** 7 in. **5.** 5 in. **6.** 6 ft **7.** 3 in.

Practice A

1. −8, 8 **2.** no square roots **3.** −11, 11

4. −1, 1 **5.** no square roots **6.** −0.4, 0.4

7. −7, 7 **8.** 0 **9.** 4 **10.** −8 **11.** −7

12. −15 **13.** −6 **14.** 12 **15.** ≈5.66

16. −1.3 **17.** 7 **18.** 5 **19.** 5 **20.** 9

21. ≈10.20 **22.** undefined **23.** ≈1.89, ≈0.51

24. ≈2.52, ≈−1.02 **25.** ≈−0.51, ≈−2.82

26. ≈5.10, ≈−0.10 **27.** ≈−6.12, ≈−1.88

28. ≈0.67, ≈−1.24 **29.** −5, 5 **30.** −9, 9

31. −7, 7 **32.** no solution **33.** $-\sqrt{12}$, $\sqrt{12}$

34. 0 **35.** 2 seconds **36.** 6 seconds

37. 7 seconds

Practice B

1. −7 **2.** −9 **3.** 0.5 **4.** ≈−6.16

5. −0.4, 0.4 **6.** 3.5 **7.** ≈−0.63, 0.63

8. 4.4 **9.** 0 **10.** 2 **11.** undefined

12. ≈3.87 **13.** undefined **14.** 11

15. ≈−15.35, ≈−0.65 **16.** ≈2.65, ≈0.35

17. ≈2.13, ≈−0.42 **18.** ≈3.41, ≈0.59

19. ≈−11.57, ≈2.57 **20.** ≈−0.62, ≈−0.88

21. −7, 7 **22.** −8, 8 **23.** 0 **24.** undefined

25. $-\sqrt{14}$, $\sqrt{14}$ **26.** $-\sqrt{5}$, $\sqrt{5}$ **27.** −5, 5

28. −2, 2 **29.** −10, 10 **30.** 3.5 cm

31. 6 seconds

Practice C

1. 12 **2.** ≈−0.71 **3.** ≈1.80 **4.** ≈−7.35

5. −0.6, 0.6 **6.** ≈6.04 **7.** ≈−0.95, ≈0.95

Lesson 9.1 *continued*

8. 8.6 **9.** 2 **10.** ≈ 13.30 **11.** 2 **12.** ≈ 6.32

13. undefined **14.** 3 **15.** $\approx -24.23, \approx 2.23$

16. $\approx 13.73, \approx -7.06$ **17.** $\approx 0.23, \approx -1.23$

18. $\approx -0.87, \approx 0.37$ **19.** $\approx -12.49, \approx 4.49$

20. $\approx 0.44, \approx -1.94$ **21.** $-10, 10$ **22.** $-4, 4$

23. $-3, 3$ **24.** $-\sqrt{\dfrac{7}{2}}, \sqrt{\dfrac{7}{2}}$ **25.** no solution

26. $-\sqrt{5}, \sqrt{5}$ **27.** $-4, 4$ **28.** $-\sqrt{2}, \sqrt{2}$

29. $-7, 7$ **30.** $-2.83, 2.83$ **31.** $-6.78, 6.78$

32. $-8.12, 8.12$ **33.** $-2.45, 2.45$

34. $-3.46, 3.46$ **35.** $-9.64, 9.64$

36. ≈ 3.87 seconds, ≈ 2.37 seconds

37. 4.24 seconds

Reteaching with Practice

1. 0.3 **2.** 6 **3.** -5 **4.** ± 10 **5.** 0 **6.** 3

7. ± 3 **8.** $\pm \sqrt{3}$ **9.** ± 7

Interdisciplinary Application

1. $r = \sqrt{\dfrac{V}{\pi h}}$ **2.** 6 cm **3.** 14 feet

4. The shorter cylinder.

5. $\dfrac{\text{Volume of 7-cm hose}}{\text{Volume of 5-cm hose}}$ $\dfrac{38.4846}{19.635} = \dfrac{1.96}{1}$

Challenge: Skills and Applications

1. $7, -3$ **2.** $-2, -8$ **3.** $7 \pm \sqrt{10}$

4. $-1 \pm \sqrt{8}$ **5.** $1 \pm \sqrt{3}$ **6.** $0, 8$

7. $-5 \pm \sqrt{5}$ **8.** $\dfrac{9}{4}, -\dfrac{15}{4}$ **9.** about 2.0 sec

10. $h = -4.9t^2 + 17$

11. $h = -4.9(t - 0.5)^2 + 20$

Lesson 9.2

Warm-up Exercises

1. 5 **2.** -10 **3.** 8 **4.** 2 **5.** 2

Daily Homework Quiz

1. ± 0.5 **2.** -1.2 **3.** 9 **4.** $-1.87, 0.37$

5. $\pm \sqrt{20}$ **6.** ± 1.53 **7.** $h = -16t^2 + 84$

Lesson Opener

Allow 20 minutes.

1–2. Equivalent expressions: $\dfrac{\sqrt{7}}{\sqrt{6}}, \sqrt{\dfrac{7}{6}}$;

$\sqrt{5} \cdot \sqrt{8}, \sqrt{5 \cdot 8}$; $\sqrt{\dfrac{10}{7}}, \dfrac{\sqrt{10}}{\sqrt{7}}$; $\sqrt{4} \cdot \sqrt{5}$,

$\sqrt{4 \cdot 5}$; $\dfrac{\sqrt{3}}{\sqrt{4}}, \sqrt{\dfrac{3}{4}}$; $\sqrt{3} \cdot \sqrt{2}, \sqrt{3 \cdot 2}$; $\sqrt{5 \cdot 7}$,

$\sqrt{5} \cdot \sqrt{7}$; $\sqrt{\dfrac{2}{5}}, \dfrac{\sqrt{2}}{\sqrt{5}}$; $\sqrt{8} \cdot \sqrt{6}, \sqrt{8 \cdot 6}$

Practice A

1. F **2.** B **3.** H **4.** A **5.** G **6.** C **7.** D

8. E **9.** $2\sqrt{5}$ **10.** $2\sqrt{3}$ **11.** $2\sqrt{10}$

12. $3\sqrt{2}$ **13.** $4\sqrt{3}$ **14.** $3\sqrt{6}$ **15.** $3\sqrt{5}$

16. $2\sqrt{3}$ **17.** $\dfrac{\sqrt{3}}{2}$ **18.** $\dfrac{3}{4}$ **19.** $\dfrac{\sqrt{3}}{5}$ **20.** $\dfrac{2\sqrt{5}}{7}$

21. $3\sqrt{2}$ **22.** $\dfrac{2\sqrt{7}}{3}$ **23.** $\sqrt{13}$ **24.** $\dfrac{5\sqrt{11}}{3}$

25. $\dfrac{4}{3}$ **26.** $2\sqrt{5}$ **27.** $\dfrac{\sqrt{2}}{3}$ **28.** $\sqrt{3}$ **29.** $\dfrac{\sqrt{10}}{2}$

30. 2 **31.** 40 **32.** 18 **33.** $4\sqrt{3} \approx 6.93$

34. $\dfrac{3}{4} = 0.75$ **35.** 18 **36.** ≈ 7.3 knots

Practice B

1. G **2.** F **3.** D **4.** B **5.** C **6.** E **7.** H

8. A **9.** $5\sqrt{2}$ **10.** $2\sqrt{5}$ **11.** $4\sqrt{15}$

12. $6\sqrt{3}$ **13.** $10\sqrt{3}$ **14.** 6 **15.** $\sqrt{5}$

16. $4\sqrt{2}$ **17.** $\dfrac{4}{5}$ **18.** $\dfrac{1}{3}$ **19.** $\dfrac{7\sqrt{3}}{4}$ **20.** $\dfrac{10\sqrt{5}}{7}$

21. $2\sqrt{5}$ **22.** $\dfrac{\sqrt{10}}{7}$ **23.** $\dfrac{5\sqrt{5}}{2}$ **24.** $\dfrac{1}{3}$

25. $\dfrac{2\sqrt{7}}{7}$ **26.** $\dfrac{\sqrt{3}}{2}$ **27.** 4 **28.** $\dfrac{1}{3}$ **29.** $\dfrac{\sqrt{14}}{3}$

30. $\dfrac{2\sqrt{7}}{5}$ **31.** $6\sqrt{42}$ **32.** $\sqrt{6}$ **33.** $2\sqrt{3}$

34. $144\sqrt{6}$ **35.** $6\sqrt{3}$ **36.** $-\sqrt{5}$

37. $2\sqrt{15} \approx 7.75$ **38.** $\dfrac{3}{8} = 0.375$ **39.** 75

Practice C

1. $4\sqrt{5}$ **2.** $7\sqrt{5}$ **3.** $4\sqrt{7}$ **4.** $12\sqrt{2}$

5. $9\sqrt{6}$ **6.** $10\sqrt{7}$ **7.** $\sqrt{5}$ **8.** $\dfrac{14\sqrt{2}}{3}$ **9.** $\dfrac{1}{2}$

Lesson 9.2 *continued*

10. $\dfrac{1}{4}$ **11.** $\dfrac{5\sqrt{5}}{3}$ **12.** $12\sqrt{3}$ **13.** $3\sqrt{3}$

14. $\dfrac{\sqrt{10}}{11}$ **15.** $\dfrac{20\sqrt{2}}{11}$ **16.** $\dfrac{4}{13}$ **17.** $\dfrac{3\sqrt{2}}{4}$

18. $\dfrac{2\sqrt{2}}{5}$ **19.** 4 **20.** $\dfrac{1}{2}$ **21.** $\dfrac{3\sqrt{7}}{7}$ **22.** $\dfrac{\sqrt{6}}{11}$

23. $20\sqrt{105}$ **24.** 2 **25.** $4\sqrt{17}$ **26.** $336\sqrt{17}$

27. $9\sqrt{2}$ **28.** $-\dfrac{21\sqrt{7}}{16}$ **29.** $3\sqrt{10} \approx 9.49$

30. $12\pi \approx 37.68$ **31.** $4\sqrt{2}\pi \approx 17.76$

32. ≈ 60.25 miles per hour

Reteaching with Practice

1. $7\sqrt{2}$ **2.** $2\sqrt{13}$ **3.** $10\sqrt{3}$ **4.** $3\sqrt{11}$

5. $\dfrac{\sqrt{11}}{2}$ **6.** $\dfrac{\sqrt{2}}{6}$ **7.** $\dfrac{\sqrt{5}}{3}$ **8.** $\dfrac{\sqrt{3}}{4}$

9. 70 meters per second

Real-Life Application

1. 35 mph **2.** 50 mph; less than 50 mph

3. about 1524.2 feet per second

Challenge: Skills and Applications

1. $13\sqrt{6}$ **2.** $17\sqrt{38}$ **3.** $\dfrac{\sqrt{3}}{2}$ **4.** $\dfrac{\sqrt{7}}{20}$

5. $30\sqrt{35}$ **6.** $420\sqrt{11}$ **7.** $\dfrac{1}{\sqrt{p}}$, or $\dfrac{\sqrt{p}}{p}$

8. $q\sqrt{pq}$ **9.** $\sqrt{13}, 5$ **10.** $7, 7$ **11.** $5, 7$

12. $3\sqrt{5}, 9$ **13.** $\sqrt{(a + b)}$ does not equal $\sqrt{a} + \sqrt{b}$, except when a, b, or both equal zero.

14. $\sqrt{a - b}$ does not equal $\sqrt{a} - \sqrt{b}$, except when b equals zero (but a does not equal zero) or when both a and b equal zero. **15.** a **16.** a

17. yes **18.** a^2 **19.** a^2 **20.** yes

Lesson 9.3

Warm-up Exercises

1. $\dfrac{4}{3}$ **2.** $\dfrac{3}{4}$ **3.** $-1, 20$ **4.** $-1, -36$

Daily Homework Quiz

1. $4\sqrt{7}$ **2.** $-6\sqrt{2}$ **3.** $4\sqrt{7}$ **4.** $\dfrac{2\sqrt{19}}{5}$

---

5. $6\sqrt{2}$ **6.** 20 units^2

Lesson Opener

Allow 10 minutes.

1. a. *Sample answer:* The graph is U-shaped and opens up. **b.** $(0, 1)$ **c.** the y-axis; the graph is symetric about the y-axis

2. a. *Sample answer:* The graph is U-shaped and opens up. **b.** $(0, 0)$ **c.** the y-axis; the graph is symetric about the y-axis

3. a. *Sample answer:* The graph is U-shaped and opens down. **b.** $(0, 2)$ **c.** the y-axis; the graph is symmetric about the y-axis

Graphing Calculator Activity

1. $(-2.6, 0), (2.6, 0)$

2. $(-2.4, 8.9), (1.8, -6.1)$ **3.** $(1.3, 5.1)$

4. none **5.** $(-2.7, 2), (0.7, 2)$ **6.** $(0.3, 0.2)$

7. 0, 1, or 2 points of intersection

Practice A

1. $a = 3, b = -5, c = 2$

2. $a = 1, b = 2, c = -3$

3. $a = -4, b = 1, c = 0$

4. $a = -1, b = 4, c = -8$

5. $a = -5, b = -1, c = 5$

6. $a = 1, b = 0, c = -4$ **7.** up, $x = 0$

8. up, $x = 1$ **9.** down, $x = -1$

10. up, $x = -1$ **11.** down, $x = 2$

12. down, $x = -\dfrac{3}{2}$ **13.** down, $x = -\dfrac{3}{4}$

14. up, $x = -1$ **15.** up, $x = -\dfrac{1}{3}$ **16.** $(0, 0)$

17. $(0, 0)$ **18.** $(0, -1)$ **19.** $(-3, -9)$

20. $(-3, -7)$ **21.** $(1, 1)$

22. vertex: $(-1, 3)$

Tables vary. Sample given.

x	-3	-2	-1	0	1
y	7	4	3	4	7

23. vertex: $(1, 4)$

Tables vary. Sample given.

x	-1	0	1	2	3
y	-8	1	4	1	-8

Lesson 9.3 *continued*

24.

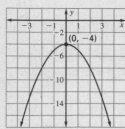

25.

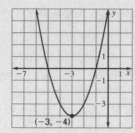

26.

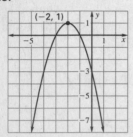

27.

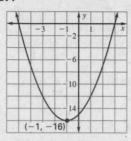

28.

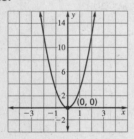

29.

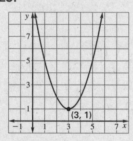

30.

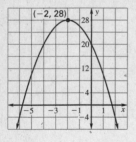

31.

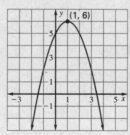

32.

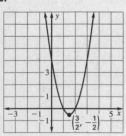

33. 630 feet

34. 1010 feet

Practice B

1. $a = -2, b = 0, c = 0$
2. $a = 1, b = -9, c = 5$
3. $a = 3, b = 7, c = 0$
4. $a = \frac{1}{2}, b = -2, c = -\frac{1}{4}$
5. $a = -4.5, b = 0, c = 4$
6. $a = 1.7, b = 2.3, c = 1.1$
7. up, $(0, 0)$, $x = 0$ **8.** down, $\left(\frac{4}{3}, \frac{16}{3}\right)$, $x = \frac{4}{3}$
9. down, $\left(-\frac{1}{2}, 9\right)$, $x = -\frac{1}{2}$ **10.** up, $(1, 1)$, $x = 1$
11. up, $(2, -14)$, $x = 2$
12. up, $\left(-\frac{3}{4}, \frac{39}{8}\right)$, $x = -\frac{3}{4}$
13. up, $\left(-\frac{7}{4}, -\frac{217}{8}\right)$, $x = -\frac{7}{4}$
14. down, $(0, -16)$, $x = 0$
15. down, $\left(\frac{7}{6}, \frac{49}{6}\right)$, $x = \frac{7}{6}$
16. vertex: $(-1, 16)$
    Tables vary. Sample given.

x	-3	-2	-1	0	1
y	12	15	16	15	12

17. vertex: $\left(-\frac{1}{3}, \frac{11}{3}\right)$
    Tables vary. Sample given.

x	-3	-2	$-\frac{1}{3}$	0	2
y	25	12	$\frac{11}{3}$	4	20

18.

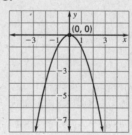

19.

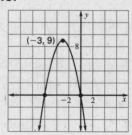

20.

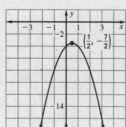

21.

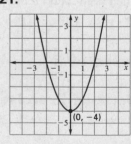

Lesson 9.3 *continued*

22.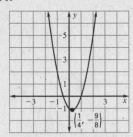

23.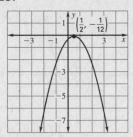

24.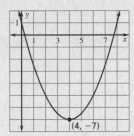

25.

26.

27. 5 feet

28. ≈ 10.3 feet   **29.** ≈ 111 units

Practice C

1. down, $(0, 0)$, $x = 0$

2. up, $\left(-\frac{1}{4}, -\frac{1}{2}\right)$, $x = -\frac{1}{4}$

3. down, $(-1, 8)$, $x = -1$

4. down, $(2, 2)$, $x = 2$ **5.** down, $(0, 2)$, $x = 0$

6. up, $\left(\frac{5}{2}, -\frac{49}{4}\right)$, $x = \frac{5}{2}$ **7.** up, $\left(\frac{5}{4}, -\frac{73}{8}\right)$, $x = \frac{5}{4}$

8. up, $\left(-\frac{3}{10}, \frac{31}{20}\right)$, $x = -\frac{3}{10}$

9. down, $\left(\frac{7}{2}, \frac{73}{4}\right)$, $x = \frac{7}{2}$

10. vertex: $(-1, -5)$

Tables vary. Sample given.

x	−3	−2	−1	0	1
y	7	−2	−5	−2	7

11. vertex: $\left(-\frac{1}{7}, -\frac{71}{7}\right)$

Tables vary. Sample given.

x	−3	−2	−$\frac{1}{7}$	0	2
y	47	14	−$\frac{71}{7}$	−10	22

12.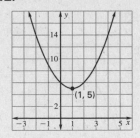

13.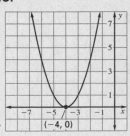

14.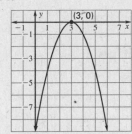

15.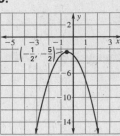

16.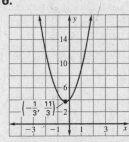

17.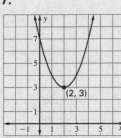

18.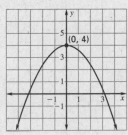

19.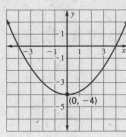

20.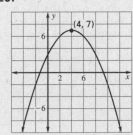

21.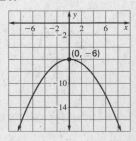

Lesson 9.3 *continued*

22.

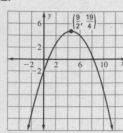

23.

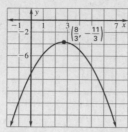

24. ≈ 3.89 feet **25.** ≈ 13.00 feet **26.** 14 feet

Reteaching with Practice

1.

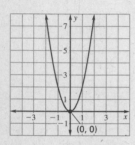

2.

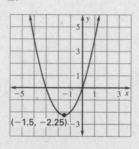

3.

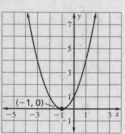

4.

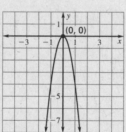

5.

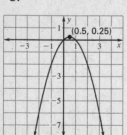

6.

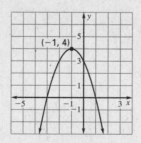

7. 25 feet

Real-Life Application

1. 2 feet **2.** 0.5 second **3.** 1.6 feet

4. 0.4 second **5.**

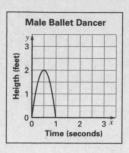

Challenge: Skills and Applications

1.

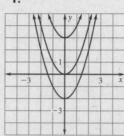

2.

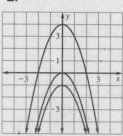

3. up k units for $k > 0$ and down $|k|$ units for $k < 0$ **4.** up k units for $k > 0$ and down $|k|$ units for $k < 0$

5.

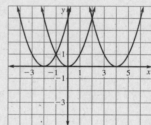

6. right h units for $h > 0$ and left $|h|$ units for $h < 0$

7. $\dfrac{8}{9}$ **8.** $\dfrac{7}{4k^2}$ **9.** $\dfrac{4}{25}$ **10.** $\dfrac{3}{2}, -5$ **11.** $-\dfrac{1}{2}, 4$

Quiz 1

1. 4.6 **2.** $-6, 6$ **3.** $\dfrac{5}{7}$ **4.** $\dfrac{2\sqrt{7}}{5}$

5. down; $(-2, 3); x = -2$ **6.**

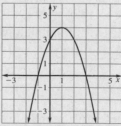

Lesson 9.4

Lesson 9.4

Warm-up Exercises

1. $5, -5$ **2.** $5, -5$ **3.** $9, -9$ **4.** $16, -16$

Daily Homework Quiz

1. up; $\left(\frac{1}{2}, \frac{13}{4}\right)$, $x = \frac{1}{2}$ **2.** down; $\left(-\frac{1}{4}, \frac{33}{8}\right)$, $x = -\frac{1}{4}$

3. 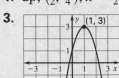 **4.** 5 ft

Lesson Opener

Allow 10 minutes.

1. B; the graph crosses the x-axis at -2 and 1.

2. D; the graph crosses the x-axis at -2 and 4.

3. A; the graph crosses the x-axis at 1 and 3.

4. B; the graph crosses the x-axis at -2 and -1.

Practice A

1. $3x^2 - 7 = 0$ **2.** $x^2 + 5x + 3 = 0$

3. $x^2 - 4x + 2 = 0$ **4.** $2x^2 - 4x - 5 = 0$

5. $x^2 - 3x = 0$ **6.** $6x^2 - 4x - 8 = 0$

7. $-1, 1$ **8.** $-2, 2$ **9.** $-2, 1$ **10.** $-3, 4$

11. $-3, 2$ **12.** $-5, 1$ **13.** $-2, 2$ **14.** $-3, 3$

15. $-6, 6$ **16.** $-8, 8$ **17.** $-5, 5$ **18.** $-7, 7$

19. $-2, 2$ **20.** $-1, 2$ **21.** $-1, 4$ **22.** $-4, 3$

23. $1, 4$ **24.** $-3, 1$ **25.** $0, 68, 104, 108, 80, 20$

26. 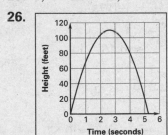 **27.** 5.25

28. 5.25 seconds

Practice B

1. $-2, 2$ **2.** $-3, 3$ **3.** $2, 5$ **4.** $-4, \frac{1}{2}$

5. $-2, \frac{5}{2}$ **6.** $-3, 5$ **7.** $-3, 3$ **8.** $-5, 5$

9. $-3, 3$ **10.** $-4, 4$ **11.** $-9, 9$ **12.** $-6, 6$

13. $-9, 9$ **14.** $-12, 12$ **15.** $-4, 4$

16. $-2, 3$ **17.** $-3, 1$ **18.** $2, 4$ **19.** $-3, -2$

20. $-\frac{1}{2}, 1$ **21.** $-5, 2$

22. $6, 8.25, 10, 11.25, 12, 12.25, 12, 11.25$

23. 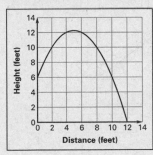 **24.** 12

25. 12 feet

Practice C

1. $-2, 1$ **2.** 3 **3.** $0, 5$ **4.** $-10, 10$

5. $-6, 6$ **6.** $-7, 7$ **7.** $-12, 12$ **8.** $-15, 15$

9. $-4, 4$ **10.** $-13, 13$ **11.** $-8, 8$

12. $-12, 12$ **13.** 1 **14.** -6 **15.** $4, -7$

16. $-1, 4$ **17.** $12, -2$ **18.** $-4, -\frac{3}{2}$ **19.** $2, 8$

20. $-\frac{3}{2}, 4$ **21.** $-2, \frac{1}{2}$ **22.** $-3, 5$ **23.** $-7, 2$

24. $-3, 4$ **25.** $-2, 1$ **26.** $-8, 6$ **27.** $-5, 4$

28. 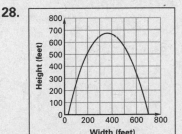 **29.** $30, 700$

30. 670 feet

Reteaching with Practice

1. ± 6 **2.** ± 4 **3.** ± 3 **4.** $-4, 3$ **5.** $2, 3$

6. $-1, 6$ **7.** $-0.63(5)^2 + 15.08(5) + 151.57$
$= 211.22 \approx 210$

Lesson 9.4 *continued*

Interdisciplinary Application

1.

2. 1999

3.
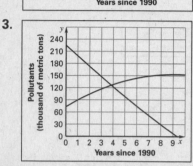

4. 1986

Challenge: Skills and Applications

1. $y = -3(x - 3)(x + 1)$

2. $y = -\frac{1}{4}(x - 3)(x - 5)$

3. $y = -5(x - 4)(x + 1)$

4. $y = \frac{1}{17}(x + 5)(x + 2)$

5. $y = \frac{1}{2}(x - 5)(x + 3)$ **6.** $y = -\frac{4}{75}x^2$

7. about 17.32 feet **8.** $y = -\frac{1}{12}x^2$

9. 48 inches higher

Lesson 9.5

Warm-up Exercises

1. 8 **2.** $2\sqrt{10}$ **3.** $2\sqrt{3}$ **4.** $2\sqrt{19}$ **5.** 5

Daily Homework Quiz

1. $-1.5, 1$ **2.** $-4, 4$ **3.** $-5, 5$ **4.** $-4, 1$
5. $-5, 2$

Lesson Opener

Allow 10 minutes.

1. a. $a = 1, b = -3, c = -10$ **b.** $-2, 5$
c. The solutions are the same. **2. a.** $a = 1$,
$b = 3, c = -4$ **b.** $-4, 1$ **c.** The solutions are
the same. **3. a.** $a = 1, b = 6, c = 8$
b. $-2, -4$ **c.** The solutions are the same.

4. *Sample answer:* Substituting the values for a, b, and c into the quadratic formula gives the solutions to the equation $ax^2 + bx + c = 0$.

Practice A

1. $4x^2 - 12 = 0$ **2.** $3x^2 + 8x + 2 = 0$

3. $x^2 - 10x + 6 = 0$ **4.** $3x^2 - 5x + 4 = 0$

5. $4x^2 + 5x = 0$ **6.** $8x^2 + 5x - 1 = 0$

7. 12 **8.** 52 **9.** 100 **10.** 12 **11.** 17

12. 32 **13.** 5, 3 **14.** $-2, -9$ **15.** $2, -7$

16. $1, -2$ **17.** $-2, \frac{1}{2}$ **18.** $1, \frac{3}{4}$ **19.** $-1, \frac{1}{4}$

20. $-2, 4$

21. $-1 + \sqrt{3} \approx 0.73, -1 - \sqrt{3} \approx -2.73$

22. $-3, -2$ **23.** $-\frac{3}{2}, 4$ **24.** $-\frac{11}{5}, 3$ **25.** $-\frac{5}{3}, \frac{1}{2}$

26. 3, 6 **27.** $\frac{4}{3}, 1$ **28.** 4 inches

29. $t = \dfrac{-95 + \sqrt{1729}}{-32} \approx 1.67$ sec.,

$t = \dfrac{-95 - \sqrt{1729}}{-32} \approx 4.27$ sec.

30. $t = \dfrac{-28 - \sqrt{1104}}{-32} \approx 1.91$ sec.

Practice B

1. 76 **2.** 41 **3.** 40 **4.** 33 **5.** 20 **6.** 134

7. $3, \frac{1}{4}$ **8.** $-3, -\frac{1}{2}$ **9.** $-5, 6$ **10.** $-4, \frac{5}{3}$

11. $-\frac{7}{4}, 2$ **12.** $\dfrac{1 + \sqrt{17}}{4} \approx 1.28$,

$\dfrac{1 - \sqrt{17}}{4} \approx -0.78$

13. $\frac{1}{2}, 1$ **14.** $\dfrac{-5 + \sqrt{35}}{2} \approx 0.46$,

$\dfrac{-5 - \sqrt{35}}{2} \approx -5.46$

15. $\dfrac{-2 + \sqrt{2}}{-6} \approx 0.10, \dfrac{-2 - \sqrt{2}}{-6} \approx 0.57$

16. $\frac{2}{5}, 1$ **17.** $\dfrac{5 + \sqrt{89}}{16} \approx 0.90$,

$\dfrac{5 - \sqrt{89}}{16} \approx -0.28$

18. $\dfrac{-9 + \sqrt{109}}{14} \approx 0.10, \dfrac{-9 - \sqrt{109}}{14} \approx -1.39$

Lesson 9.5 *continued*

19. $-4, 2$ **20.** $-\frac{1}{2}, 3$ **21.** $-\frac{4}{3}, \frac{3}{2}$

22. $\dfrac{-5 + \sqrt{13}}{6} \approx -0.23, \dfrac{-5 - \sqrt{13}}{6} \approx -1.43$

23. $\dfrac{-2 + \sqrt{14}}{-5} \approx -0.35, \dfrac{-2 - \sqrt{14}}{-5} \approx 1.15$

24. $\dfrac{7 + \sqrt{41}}{-4} \approx -3.35, \dfrac{7 - \sqrt{41}}{-4} \approx -0.15$

25. $t = \dfrac{-20 - \sqrt{432}}{-32} \approx 1.27$ sec

26. $t = \dfrac{-125 - \sqrt{15817}}{-32} \approx 7.84$ sec

27. $t = \sqrt{0.75} \approx 0.87$ sec

28. $t = \dfrac{-15 - \sqrt{993}}{-32} \approx 1.45$ sec

Practice C

1. 40 **2.** 32 **3.** 9 **4.** 304 **5.** 177 **6.** 208

7. $2, -1$ **8.** $1, \frac{1}{2}$ **9.** $-1, \frac{2}{3}$ **10.** $-\frac{1}{3}, -\frac{1}{5}$

11. $\frac{1}{2}, -\frac{15}{4}$

12. $\dfrac{9 + \sqrt{117}}{18} \approx 1.10, \dfrac{9 - \sqrt{117}}{18} \approx -0.10$

13. $\dfrac{3 + \sqrt{6}}{3} \approx 1.82, \dfrac{3 - \sqrt{6}}{3} \approx 0.18$

14. $\dfrac{-7 + \sqrt{129}}{8} \approx 0.54, \dfrac{-7 - \sqrt{129}}{8} \approx -2.29$

15. $\dfrac{19 + \sqrt{409}}{4} \approx 9.81, \dfrac{19 - \sqrt{409}}{4} \approx -0.31$

16. $\dfrac{5 + \sqrt{61}}{-18} \approx -0.71, \dfrac{5 - \sqrt{61}}{-18} \approx 0.16$

17. $\dfrac{-3 + \sqrt{37}}{4} \approx 0.77, \dfrac{-3 - \sqrt{37}}{4} \approx -2.27$

18. $\dfrac{-8 + \sqrt{58}}{6} \approx -0.06,$ **19.** $-5, 10$

$\dfrac{-8 - \sqrt{58}}{6} \approx -2.60$

20. $\dfrac{9 + \sqrt{17}}{4} \approx 3.28, \dfrac{9 - \sqrt{17}}{4} \approx 1.22$

21. $\dfrac{9 + \sqrt{193}}{8} \approx 2.86, \dfrac{9 - \sqrt{193}}{8} \approx -0.61$

22. $\dfrac{7 + \sqrt{137}}{-4} \approx -4.68, \dfrac{7 - \sqrt{137}}{-4} \approx 1.18$

23. $\dfrac{-25 + 2\sqrt{155}}{5} \approx -0.02,$

$\dfrac{-25 - 2\sqrt{155}}{5} \approx -9.98$

24. $\dfrac{9 + \sqrt{87}}{2} \approx 9.16, \dfrac{9 - \sqrt{87}}{2} \approx -0.16$

25. $-4 + 2\sqrt{11} \approx 2.63$

26. $t = \dfrac{30 + \sqrt{836}}{32} \approx 1.84$ sec **27.** 1995

28. $t = \sqrt{1.75} \approx 1.32$ sec

29. Answers will vary.

Reteaching with Practice

1. $1, 3$ **2.** $-5, -4$ **3.** $-3, 2$

4. 1.25 seconds

Interdisciplinary Application

1. $60 = 110I - 30I^2$

2. $30I^2 - 110I + 60 = 0$

$a = 30; b = -110; c = 60$

3. $I = \frac{2}{3}$ amperes

4. $30I^2 - 110I + 60 = 0$

$30\left(\frac{2}{3}\right)^2 - 110\left(\frac{2}{3}\right) + 60 \overset{?}{=} 0$

$\frac{40}{3} - \frac{220}{3} + 60 \overset{?}{=} 0$

$0 = 0$

5. $I = \frac{35}{37}$ amperes

Challenge: Skills and Applications

1, 2. Accept equivalent equations.

1. $3x^2 - 7x - 5 = 0$

2. $-1.2x^2 + 4.1x + 0.7 = 0$ **3.** $x = \frac{3}{4}$

4. -0.3 sec, 1.8 sec **5.** The ball hits the ground after being thrown, that is, when $t > 0$.

Lesson 9.6

Warm-up Exercises

1. 56 **2.** 0 **3.** -111 **4.** 16 **5.** 36

Lesson 9.6 *continued*

Daily Homework Quiz

1. 1 **2.** $-\frac{3}{2}, \frac{5}{3}$ **3.** $6x^2 - 17x + 12 = 0; \frac{4}{3}, \frac{3}{2}$

4. $\frac{5 - \sqrt{17}}{4} \approx 0.22, \frac{5 + \sqrt{17}}{4} \approx 2.28$

5. $-2, 2$ **6.** 3.75 sec

Lesson Opener

Allow 15 minutes.

1.

Quadratic Equation	Is $b^2 - 4ac$ >, =, or < 0?
$x^2 + 3x - 4 = 0$	>
$2x^2 + x - 5 = 0$	>
$-4x^2 + 2x + 1 = 0$	>
$-x^2 - 4x - 4 = 0$	=
$x^2 + 2x + 1 = 0$	=
$3x^2 + x + 5 = 0$	<
$-2x^2 + 2x - 1 = 0$	<
$x^2 + 4x + 6 = 0$	<

Quadratic Equation	Solution(s) of Equation	Number of Solutions
$x^2 + 3x - 4 = 0$	$-4, 1$	2
$2x^2 + x - 5 = 0$	$-1.85, 1.35$	2
$-4x^2 + 2x + 1 = 0$	$-0.31, 0.81$	2
$-x^2 - 4x - 4 = 0$	-2	1
$x^2 + 2x + 1 = 0$	-1	1
$3x^2 + x + 5 = 0$	none	0
$-2x^2 + 2x - 1 = 0$	none	0
$x^2 + 4x + 6 = 0$	none	0

2. Sample answer: if $b^2 - 4ac > 0$, there are two solutions. If $b^2 - 4ac = 0$, there is one solution. If $b^2 - 4ac < 0$, there is no real solution.

Practice A

1. C **2.** A **3.** B **4.** none **5.** 1 **6.** 2

7. 1 **8.** 1 **9.** none **10.** 2 **11.** 2 **12.** 2

13. none **14.** 2 **15.** none **16.** none

17. yes, ≈ 4.37 **18.** D **19.** yes

Practice B

1. 1 **2.** 2 **3.** 2 **4.** none **5.** 2 **6.** 2

7. none **8.** 1 **9.** 1 **10.** 2 **11.** 2

12. none **13.** no **14.** yes **15.** 2002

16. yes **17.** yes **18.** no **19.** yes **20.** no

Practice C

1. 2 **2.** 2 **3.** none **4.** 1 **5.** none **6.** 2

7. none **8.** 1 **9.** 2 **10.** 2 **11.** none **12.** 1

13–15. Check graphs. **13.** The equation has two solutions for all values of $c < 9$, one solution when $c = 9$, and no real solution for $c > 9$.
14. The equation has two solutions for all values of $c < 2$, one solution when $c = 2$, and no real solution for $c > 2$. **15.** The equation has two solutions for all values of $c < 3$, one solution when $c = 3$, and no real solution for $c > 3$.
16. 2000 **17.** 2010 **18.** no **19.** yes

Reteaching with Practice

1. one solution **2.** two solutions

3. no real solution **4.** one solution

5. no real solution **6.** two solutions

7. The value of the discriminant is 701.1084, so the company's revenue will reach $150 million.

8.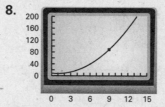
From the graph, you can see that the revenue will reach $90 million in about 9 years.

Real-Life Application

1. The equation gives S in millions of dollars.

2. $5.4t^2 + 8.3t - 377.5 = 0$

3. $b^2 - 4ac = 8222.89$ Because the discriminant is positive, the equation has two solutions. The business will not be risky.

4. $-5.3t^2 + 34.25t - 58.6 = 0$

5. $b^2 - 4ac = -69.2575$ Because the discriminant is negative, the equation has no solution. The business will probably not succeed.

Lesson 9.6 *continued*

Math and History

1. 2 solutions; 1 reasonable solution; accept reasonable responses **2.** 30 **3.** A good answer should include the fact that Babylonian algebra was used in practical, real-life situations, not abstract theories; accept reasonable responses

Challenge: Skills and Applications

1. $3, -3$ **2.** $\frac{25}{3}$ **3.** $\frac{36}{5}, -\frac{36}{5}$ **4.** $2, -2$
5. $-\frac{4}{25}$ **6.** $\frac{1}{3}, -\frac{1}{3}$ **7.** $k > -\frac{9}{2}$ **8.** $k < \frac{25}{16}$
9. $k > -\frac{1}{10}$ **10.** $k < \frac{49}{60}$ **11. a.** $4q^2 - 4p^2$
b. 2 **12. a.** $q^2 + 4p^2$ **b.** 2 **13. a.** $p^2 - 4q^2$
b. 0 **14. a.** $4p^2 - 4p^2$ **b.** 1

Quiz 2

1. $-10, 10$ **2.** $2, -1$ **3.** $6, -3$
4. one solution **5.** $c < 1$ **6.** 1.85 seconds

Lesson 9.7

Warm-up Exercises

1. yes **2.** no **3.** no **4.** yes **5.** yes

Daily Homework Quiz

1. no real solution **2.** two solutions
3. $\frac{1}{4}$; yes **4.** $c > -4, c = -4, c < -4$
5. yes

Lesson Opener

Allow 10 minutes.

1. D; the area, x^2, must be greater than or equal to 36 ft². **2.** A; the area, $(x + 5)(x + 6)$, is to be greater than 35 ft². **3.** B; the area, $(10 - 2x)(12 - 2x)$, must be less than or equal to 99 ft².

Practice A

1. no **2.** no **3.** yes **4.** yes **5.** yes **6.** no
7. yes **8.** no **9.** yes **10.** C **11.** A **12.** E
13. B **14.** F **15.** D

16.
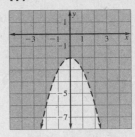

17.
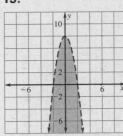

18.
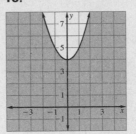

19.
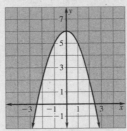

20.
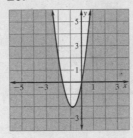

21.
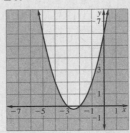

22.
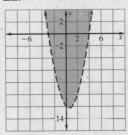

23.
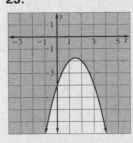

24.
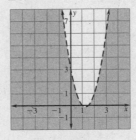

25. Less; from the graph, 402 billion gallons for 1995 lies below the parabola.

Practice B

1. no **2.** no **3.** yes **4.** yes **5.** yes **6.** yes

Lesson 9.7 *continued*

7. C **8.** A **9.** E **10.** B **11.** F **12.** D

13.

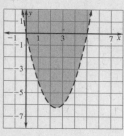

14.

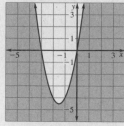

15.

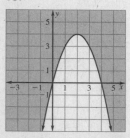

16.

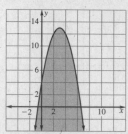

17.

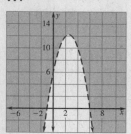

18.

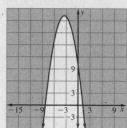

19.

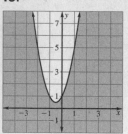

20.

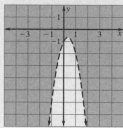

21.

22.

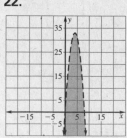

23.

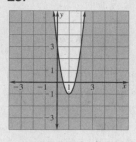

24.

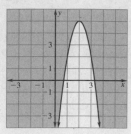

25. 10 cm

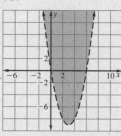

26. yes

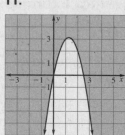

Practice C

1. no **2.** no **3.** yes **4.** yes **5.** yes **6.** no
7. C **8.** A **9.** B

10.

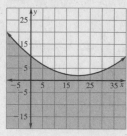

11.

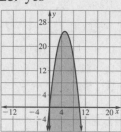

12.

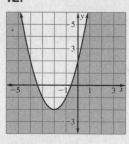

13.

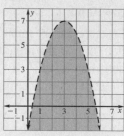

14.

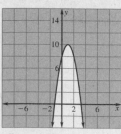

15.

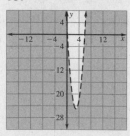

Lesson 9.7 *continued*

16.

17.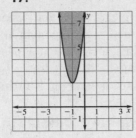

4.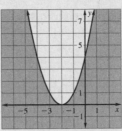

5.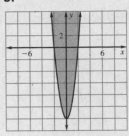

18.

19.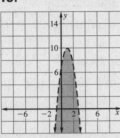

6.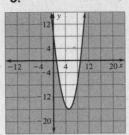

7. Answers will vary. *Sample answer:* Two possible scenarios are a width of 10 feet and a length of 14 feet or a width of 16 feet and a length of 20 feet.

20.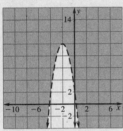

21.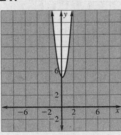

8. $x^2 + 10x - 96 > 0$

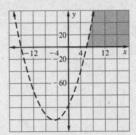

So, the width should be greater than 6 feet and the length greater than 16 feet.

22. $L = 90 - x$; $W = 60 - x$

23. $(90 - x)(60 - x)$

24.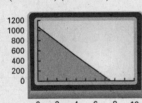

25. no, yes

26. between 0 and 20 meters

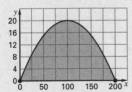

Real-Life Application

1.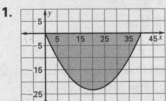

$y \geq 0.0581x^2 - 2.3238x$

$y \leq 0$

2. Yes; At its deepest point the river is about 23 feet deep.

3. No; When you are 8 feet from the bank of the river, the river is only about 15 feet deep.

Reteaching with Practice

1. $(2, 0)$ is a solution **2.** $(1, -1)$ is a solution

3. $(2, -3)$ is not a solution

Lesson 9.7 *continued*

Challenge: Skills and Applications

1.

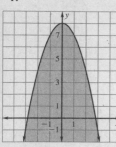

2.

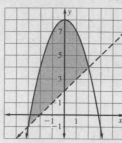

3. no   **4.** yes   **5.** no   **6.** no

7.

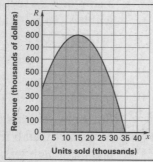

8. yes   **9.** yes

10.

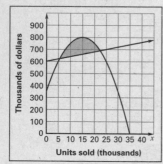

11. The values in the shaded region indicate when revenue could exceed cost so that the company could make a profit, that is, when R could be greater than C.

Lesson 9.8

Warm-up Exercises

1. 4   **2.** 4   **3.** 3   **4.** 0.95

Daily Homework Quiz

1. yes   **2.** C   **3.** 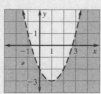   **4.** 59.5 ft

Lesson Opener

Allow 10 minutes.

1. linear; when you connect the points, a line is formed.   **2.** exponential; when you connect the points, a curve that appears to be exponential is formed.   **3.** quadratic; when you connect the points, a parabola is formed.

Graphing Calculator Activity

1. linear; $y = 0.556x - 17.778$

2. quadratic; $y = -x^2 + 24x - 60$

3. exponential; $y = 14.957(0.960)^x$

Practice A

1. exponential   **2.** quadratic   **3.** linear

4. linear   **5.** exponential   **6.** quadratic

7.

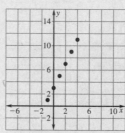

Linear

8.

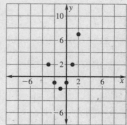

Quadratic

9.

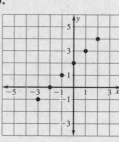

Linear

10.

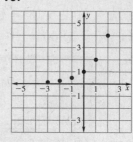

Exponential

11.

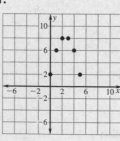

Quadratic

12.

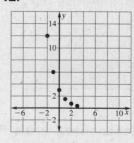

Exponential

Lesson 9.8 *continued*

13. C   **14.** Answers may vary. *Sample answer:* CDs became more popular.

15. linear; $B = 100 + 5t$ **16.** $A = 6s^2$

Practice B

In Exercises 1–3, answers will vary. Examples are given.

1. $y = 4x - 5$ **2.** $y = 4(1.05)^x$

3. $y = 3x^2 - x$

4.

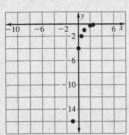

Exponential

5.

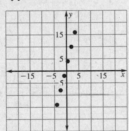

Quadratic

6.

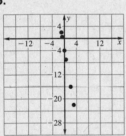

Linear

7.

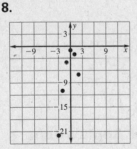

Linear

8.

Quadratic

9.

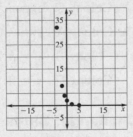

Exponential

10. quadratic, $E = 5v^2$

11.

Year t	Cell Sites C	Change in C from previous year
0	5784.22	—
1	7623.60	1839.38
2	10,047.91	2424.31
3	13,243.14	3195.23
4	17,454.46	4211.32
5	23,004.98	5550.52

12. The change is increasing exponentially.

Practice C

1.

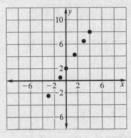

Linear

2.

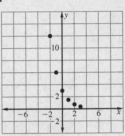

Exponential

3.

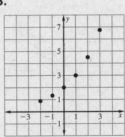

Exponential

4.

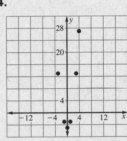

Quadratic

5.

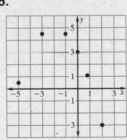

Quadratic

6.

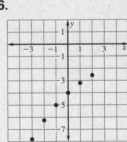

Exponential

7. $V = 500 - 50t$

Answers

Lesson 9.8 *continued*

8.

Year t	Population P	Year t	Population P
0	77,995,800	50	156,100,800
10	90,016,800	60	177,121,800
20	103,837,800	70	199,942,800
30	119,458,800	80	224,563,800
40	136,879,800	90	250,984,800

9. The population change fits a linear model.

10. absolute value

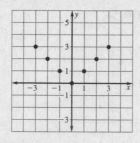

Reteaching with Practice

1. quadratic

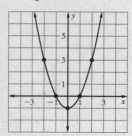

2. linear

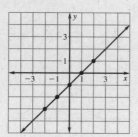

3. exponential

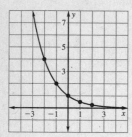

4. quadratic

Cooperative Learning Activity

1. Answers will vary.   2. Answers will vary.

Interdisciplinary Application

1.

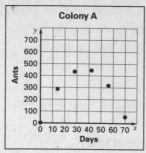

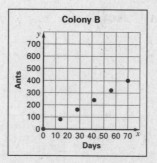

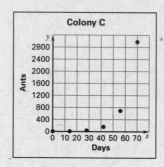

2. Colony A: Quadratic

Colony B: Linear

Colony C: Exponential

3. Colony A: $y = -0.35x^2 + 25.24x + 2$

Colony B: $y = 5.7x + 2$

Colony C: $y = 2(1 + 0.11)^x$

Challenge: Skills and Applications

1. *Sample answer:* 6

2. *Sample answer:* $y = x + 1$

3. *Sample answer:* 9

4. *Sample answer:* $y = \frac{1}{3}x^2 + 1$

5, 6. Sample answers: value of k is 16 and model is $y = 2^{x-1}$; value of k is 10 and model is $y = 1.6^x$

7. $y = 7.576(1.11)^x$

Lesson 9.8 *continued*

8. 12.766; the estimate with the model is much lower than the actual data value.

9. $y = 0.833x^2 + 7.576$ **10.** 28.401; the estimate with the model is a little lower than the actual data value, but it is fairly close.

11. quadratic

Review and Assessment

Review Games and Activities

Across   1. Quotient property   **5.** Roots

6. Square roots   **8.** Discriminant   **9.** Leading coefficient   **10.** Parabola   **13.** Radical

Down   2. Exponential model   **3.** Quadratic formula   **4.** Quadratic   **6.** Simplest form

7. Linear model   **11.** Radicand   **12.** Vertex

Test A

1. 9   **2.** -10 **3.** 5   **4.** 4   **5.** $x = -9$ and $x = 9$ **6.** $x = -7$ and $x = 7$ **7.** $3\sqrt{5}$

8. $3\sqrt{6}$ **9.** $\dfrac{4}{5}$ **10.** $\dfrac{\sqrt{5}}{4}$

11.

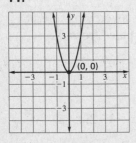

12.

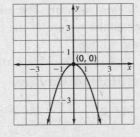

13. $x = -2$ and $x = 4$ **14.** $x = -2$ and $x = 2$
15. $x = -5$ and $x = 4$ **16.** $x = 2$ and $x = 3$
17. 1 and 2   **18.** -1 and 1   **19.** one solution
20. no solution

21.

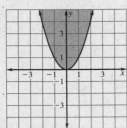

22.

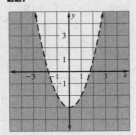

23. 50,000 units   **24.** exponential   **25.** linear

26. quadratic

Test B

1. 13   **2.** -25 **3.** 3   **4.** $x = -4$ and $x = 4$

5. $x = -6$ and $x = 6$ **6.** $\dfrac{3}{4}$ **7.** $\dfrac{2}{5}$

8.

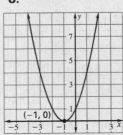

9.

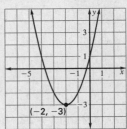

10. 2 seconds   **11.** $x = -2$ and $x = 3$

12. $x = 2$ and $x = 4$ **13.** $x = -3$ and $x = 5$

14. $x = -6$ and $x = -4$ **15.** 5 and 7

16. 1 and 6   **17.** one solution   **18.** no solutions

19.

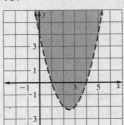

20.

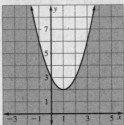

21. 100,000 units   **22.** exponential   **23.** linear

24. quadratic

Test C

1. 0.25   **2.** ± 1.2 **3.** $2\sqrt{15}$

4. $x = -3$ and $x = 3$ **5.** $x = -3$ and $x = 3$

6. $\dfrac{2\sqrt{3}}{5}$ **7.** $\dfrac{8}{5}$

8.

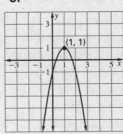

9.

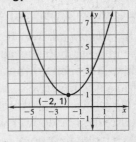

10. 3 seconds   **11.** $x = -5$ and $x = 2$

Review and Assessment *continued*

12. $x = 3$ **13.** $x = 3$ and $x = 5$ **14.** $x = -6$
and $x = 4$ **15.** -7 and 5 **16.** -8 and 6

17. one solution **18.** no solutions

19. 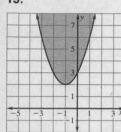 **20.**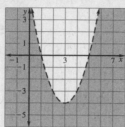

21. 150,000 units **22.** exponential **23.** linear

24. quadratic

SAT/ACT Chapter Test

1. C **2.** B **3.** A **4.** D **5.** A **6.** A **7.** A
8. D **9.** C

Alternative Assessment

1. a. Answers may vary; *Sample answer:* A
penny is dropped from the top of a 400 story
building. How long will it take to hit the ground?
A penny is thrown down at 10 feet per second.
This time, how soon will it hit the ground? A
model rocket is shot vertically into the air from a
height of 10 feet? How high is it after 2 seconds?
b. Complete answers should address these points:
• The model $h = -16t^2 + s$ is used for a
dropped object and $h = -16t^2 + vt + s$ for
objects thrown down or up. • When finding how
long it takes an object to hit the ground, 0 is
substituted in for h, since the height at ground
level is zero. • Heights chosen other than the
ground should be set equal to zero before solving.
• Initial height is substituted in for s, and the
initial velocity for v. • Initial velocity is positive
for an object thrown upward and negative for an
object thrown downward.

2. a. *Sample answer:* the graph would be the
shape of a parabola, because the model is quadratic. It would open downward because the leading
coefficient is negative. **b.** Check graphs.
Sample answer: As revolutions per minute
increase, the horsepower will increase until it
reaches its highest point, after which horsepower
decreases. **c.** 857; substitute 8 in for r in the
equation for horsepower and simplify.

d. Estimates should be close to 8248 rev/min and
9307 rev/min. *Sample answer:* to solve on a
graphing calculator, use the calculator's trace
feature to see when H is close to 876; to solve
algebraically substitute 876 in for H, set equation
equal to zero, and solve using the quadratic
formula. **3.** 1%; *Sample answer:* Substitute the
r-coordinate into the equation to find the first
maximum horsepower to be 893; $\frac{902}{893} \approx 1.01$, a 1%
improvement.

Project: Light Square

1. The lighted area will depend on the size of the
hole. Students may need to interpolate to find a
square. *Sample answer:* about 289 squares

2. *Sample answer:* about 1089 squares

3. Make sure students solve the equation $y = kd^2$
for k when y equals their answer to Exercise 1 and
$d = 6$ and then when y equals their answer to
Exercise 2 and $d = 12$. They should then take the
average of the k-values found to use in the final
equation. *Sample answer:* $y = 7.8d^2$

4. Check that predictions are the y-value from the
equation in Exercise 3 when $d = 9$.
Sample answer: about 632 squares; the prediction
should be close.

5. *Sample answer:* The area covered by a light
varies as the square of the distance from the
limited source.

Cumulative Review

1. 54 **2.** $-\dfrac{64t^2}{81}$ **3.** $-10x$ **4.** $2w$ **5.** $-\dfrac{y}{2}$

6. 1 **7.** -4.16 **8.** -1.98 **9.** -2.01

10. 2.36 **11.** $m = \frac{2}{15}, b = -\frac{8}{5}$

12. $m = -\frac{1}{3}, b = \frac{3}{2}$ **13.** $m = -\frac{9}{10}, b = \frac{1}{5}$

14. $m = \frac{2}{11}, b = \frac{70}{11}$ **15.** $m = \frac{16}{21}, b = 0$

16. $m = 0, b = 4.88$ **17.** $y = -\frac{1}{2}x + 2$

18. $y = -\frac{9}{5}x + \frac{8}{5}$ **19.** $y = -\frac{7}{5}x - \frac{2}{5}$

20. $y = -\frac{1}{11}x - \frac{10}{11}$ **21.** $8x - y = 19$

22. $5x + 5y = 61$ **23.** $72x + 8y = 19$

24. $n < 3$

25. $x \geq 2$

Review and Assessment *continued*

26. $a > \frac{3}{4}$

27. $-3.5 \le y$

28. $0 < c \le \frac{3}{4}$

29. $1.9 < x$

30. $(4, -8)$ **31.** $\left(-\frac{3}{4}, \frac{1}{2}\right)$ **32.** $\left(\frac{13}{3}, -2\right)$

33. 78 **34.** $-3888x^{17}$ **35.** $16x^6y^6$

36. $-r^{13}s^{19}$ **37.** $a^9b^{11}c^7$ **38.** $-432xy^{23}$

39. -11 **40.** 3.87 **41.** 17 **42.** 0.14

43. ± 0.03 **44.** ± 4.5 **45.** $\dfrac{\sqrt{3}}{4}$ **46.** $\dfrac{2\sqrt{2}}{5}$

47. 33 **48.** $\dfrac{\sqrt{46}}{9}$ **49.** 2

50. $-\dfrac{4\sqrt{5}}{15}$ **51.** opens up, $x = \frac{2}{5}$

52. opens up, $x = \frac{63}{2}$

53. opens down, $x = -\frac{5}{252}$

54. $x = \pm\sqrt{6}$ **55.** $x = \pm 12$ **56.** $x = \pm 10$

57. $-1, 3$ **58.** $-1 \pm \sqrt{5}$ **59.** $\dfrac{-1 \pm \sqrt{6}}{5}$

60. $\dfrac{1}{3}, 2$ **61.** $\dfrac{-3 \pm \sqrt{21}}{2}$ **62.** $\dfrac{-5 \pm \sqrt{217}}{8}$

63. two solutions   **64.** no solution

65. one solution

66.

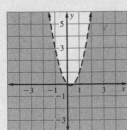

67.

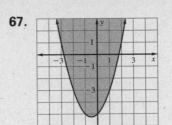

68.

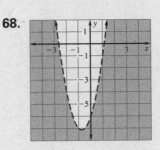

69. Exponential   **70.** Quadratic

71. Quadratic   **72.** Linear